Samuel G. Love, Mary E. Willard

**Industrial Education**

A Guide to Manual Training

Samuel G. Love, Mary E. Willard
**Industrial Education**
*A Guide to Manual Training*
ISBN/EAN: 9783743465169

Manufactured in Europe, USA, Canada, Australia, Japa

Cover: Foto ©berggeist007 / pixelio.de

Manufactured and distributed by brebook publishing software (www.brebook.com)

Samuel G. Love, Mary E. Willard

**Industrial Education**

# Industrial Education.

## A Guide to Manual Training

BY

SAMUEL G. LOVE,

SUPERINTENDENT OF THE JAMESTOWN, N. Y. PUBLIC SCHOOLS,

ASSISTED BY

MARY R. WILLARD,

With Many Illustrations

By VESTA WILLARD,

NEW YORK AND CHICAGO.

E. L. KELLOGG & CO.,

1889.

Copyright, 1887,
F. L. KELLOGG & CO.,
New York.

# PREFACE.

The lessons in this volume were originally prepared as a guide for the instructors and pupils in the Industrial department of the schools with which the authors are connected. A complete system of Manual Training having been organized and put in operation, it became necessary that it should be formulated for the convenience of the classes, and for the guidance of the instructors in charge of them.

It was not supposed that there would be a demand for the publication of this course in Manual Training. But the subject of Industrial Education has been rapidly growing in the minds of teachers and others interested in the development and growth of educational methods. Inquiries came with increased frequency as to what was being done in the Jamestown schools, and whether it was practicable to make Manual Training a feature of education in the public schools. It was thought best to make no reply until actual tests, in all the grades from the Primary to the High Schools, had been made.

We could now give to the public a statement of our experiences and the methods we had adopted for conducting a system of Manual Training in our schools, but we hesitated because of grave doubts of our ability properly to present the matter. Messrs. E. L. Kellogg & Co., the educational publishers, have watched our course for years and encouraged us, and at their request we at length consented, and prepared the pages that follow.

It should be remembered that the field of labor in preparing such a volume is new and mostly untried. It could not be hoped to overcome difficulties in the way of making it acceptable in a scientific and literary point of view.

It is no more than a plain, unvarnished explanation of the way in which Manual Training has been introduced and carried on in the public schools of Jamestown for several successive years. And it may encourage if we add that Manual Training has grown in favor with both pupils and patrons here from the very beginning.

With what success the work of preparing this volume has been accomplished must be left for those to decide who undertake its use. Let it be used in the same spirit in which it is given to the public— an earnest desire to find better methods of educating the children.

Many and grateful acknowledgments are due to those who have given encouragement and aided in the preparation of the text for the publishers.

S. G. Love.

Jamestown, N. Y., May, 1887.

# CONTENTS.

| Chap. | Page |
|---|---|
| Preface, | iii–iv |
| Contents, | v–vi |
| Suggestions to Teachers, | vii |
| How to Introduce Manual Training, | viii |
| Manual Work for Girls of the Grammar School, | xii |
| Manual Training for Boys in the Grammar School. | xv |
| Manual Training in the High School, | xviii |

## PART I.

| | Page |
|---|---|
| The Claims of Manual Training, | 1–19 |
| I. Upon what Grounds is the Claim made for the Introduction of Manual Training in our Schools?. | 1 |
| II. Is it Practicable to Add Manual Training to the Present Course? | 13 |
| III. How can Manual Training be Given in our Schools, | 14 |
| Manual Training in the Primary Schools, | 15 |
| " " " Second Primary Schools, . | 16 |
| " " " Grammar Schools, | 16 |
| " " " High Schools, | 17 |
| Time required for Instruction in Manual Training, | 18 |
| Instructions for Manual Training, | 18 |
| Expense of Manual Training, | 19 |

## PART II.

| Chap. | | Page |
|---|---|---|
| I. | Introduction of Industrial or Manual Training into the Jamestown Public Schools, | 20–29 |
| II. | Course of Study of the Jamestown Public Schools, Adopted June, 1886, | 30–38 |

## PART III.

| | | |
|---|---|---|
| I. | Suggestions, Lessons, and Methods of Instructions in Manual Training in the Primary School—First Grade, | 39 |
| | Block Building, | 40 |
| | Straw Stringing, | 70 |
| | Bead Stringing, | 80 |
| | Learning Colors, | 83 |
| | Tablet Laying, | 90 |
| | Paper Folding, | 95 |
| | Writing, Drawing, Gymnastics, and Review Lessons, | 103 |
| II. | The Primary School—Second Grade, | 106 |
| | Stick Laying, | 107 |
| | Picture Cutting, | 111 |
| | Scrap-book Making, | 113 |
| | Spool Work, | 114 |
| | Paper Embroidery, | 118 |
| | Braiding, | 122 |
| | Writing, Drawing, and Gymnastics, | 124 |
| III. | The Primary School—Third Grade, | 125 |
| | Perforated Cardboard Embroidery, | 126 |
| | Star Plaiting, | 128 |
| | Mat Weaving, | 132 |
| | Writing, Drawing, and Gymnastics, | 138 |
| IV. | The Primary School—Fourth Grade, | 140 |
| | Slat Plaiting (*Advanced*), | 141 |
| | Crocheting Chain Stitch, | 142 |
| | Paper Folding (*Advanced*), | 145 |

| CHAP. | | PAGE |
|---|---|---|
| | Perforated Cardboard Embroidery (*Advanced*), | 150 |
| | Penmanship, Drawing, and Gymnastics, | 151 |
| V. | THE PRIMARY SCHOOL—FIFTH GRADE, | 153 |
| | Sewing over-and-over, | 154 |
| | Crocheting, | 156 |
| | Paper Folding and Mounting, | 160 |
| | Penmanship, Drawing, and Gymnastics, | 164 |
| VI. | THE PRIMARY SCHOOL—SIXTH GRADE, | 166 |
| | Hemming, | 167 |
| | Pease Work, | 169 |
| | Knitting, | 170 |
| | Tissue-paper Flowers, | 176 |
| | Penmanship, Drawing, and Gymnastics, | 184 |

## PART IV.

| | | |
|---|---|---|
| I. | MANUAL WORK FOR BOYS AND GIRLS—INTRODUCTORY, | 186 |
| II. | THE GRAMMAR SCHOOL—SEVENTH, EIGHTH AND NINTH GRADES, | 189 |
| III. | LESSONS IN PLAIN SEWING, RUNNING, GATHERING, STITCHING, AND OVERCASTING, | 191 |
| | Crocheting, | 194 |
| | Knitting, | 195 |
| | Mending, Patching and Darning, and making Button-holes, | 195 |
| IV. | THE SHOP FOR WOOD-WORK—INTRODUCTORY, | 198 |
| | The Shop, | 199 |
| | The Hammer, | 201 |
| | Material, | 202–206 |
| | Drawing Lines and Laying off Distances, | 206 |
| | Sawing, | 210 |
| | Planing, | 213 |
| | Use of Hammer, Saw, Plane and Marking Tools, | 220 |
| | Boring, | 225 |
| | Chiseling, and the Mortise and Tenon Joint, | 229 |
| | Miter Joint, Dowel Joint, and Dovetail Joint, | 235 |
| | Penmanship, Drawing, and Gymnastics, | 249 |

## PART V.

| Chap. | | Page |
|---|---|---|
| I. | Manual Training in the High School—Introductory, | 251 |
| II. | Cutting—The Sewing Machine, | 254 |
| | Embroidery, | 255 |
| III. | Cooking—The Kitchen—Materials—Plan of Work, | 258 |
| IV. | Foot Power Machine Tools, | 263 |
| | The Scroll Saw, | 271 |
| | The Circular Saw, | 274 |
| V. | Finishing, | 275 |
| VI. | Drawing and Construction, | 280 |
| VII. | Printing, | 286 |
| Index, | | 296 |

# SUGGESTIONS TO TEACHERS.

THIS volume, as the title indicates, is intended as a guide to Manual Training. It is not meant to dictate a set of cast-iron methods, but solely to enlighten the judgment of the teacher who undertakes to use it. It will doubtless fall into the hands of those whose ability or experience will qualify them to amend or improve some of its features; the purposes of its preparation will be greatly served by the addition of these amendments and improvements.

A few suggestions, in part explanatory of its use, are offered.

1. The book should be carefully read, before any attempt is made to use it as a text-book.

2. The teacher should become familiar with the theory, and as much as possible with the practice of each lesson, before undertaking to give the instruction to the pupils.

3. It is not necessary to put Manual Training in all the grades or classes at once. One or perhaps two classes that can be cared for the most conveniently may undertake it at first, leaving it to be introduced to the others as circumstances may permit. As the difficulties in the way of introducing it in the primary grades are much less than in other grades, it is well to begin with them.

4. Although results may not be quite as apparent, it should be remembered that elementary Manual

Training is as useful and quite as important as elementary training in Reading or Numbers.

5. In all the primary grades the lessons in Manual Training should be put in the *program for regular work, as a daily exercise;* or, at all events, so as to come up on alternate days.

6. If the value of Manual Training to the pupil is felt by the teacher to be as great as any of the subjects of study and practice, the instructor will be less apt to lack the confidence necessary to make it a success.

7. It is believed that money to defray the expenses will, at an early day, be provided by law; in the mean time, it may be raised by entertainments given by the students and pupils, or in other ways which will suggest themselves to teachers and patrons, whose spirit and purpose will not allow them to wait.

## HOW TO INTRODUCE MANUAL TRAINING.

Those teachers who desire to introduce Manual Training, should note carefully the following directions:—

*First.* Make the *first attempt* at introduction by placing it in the Primary Grades, and only one or two occupations at a time. When the use of a little has created in pupil and patron a desire for more, ways and means for widening the sphere of action will not be wanting.

*Second.* *The teacher* must have some inspiration upon the subject; must arouse interest in the class by the contagion of his own enthusiasm; must himself show belief in the work, and awaken it in the workers.

As stated several times in the pages which follow, he may combine the Industrial work with that of the ordinary school routine, until such a time as it may be deemed expedient to procure an instructor for Manual work alone. In whichever of the aforenamed capacities he **works**, it is highly essential that he be thoroughly acquainted with the specific details of each occupation. Preparation must precede teaching in Manual no less than in Mental Training.

*Third.* *The room and time* may be the same as those usually occupied for recitations. With young children, an exercise in Manual work ought never to occupy more than fifteen or twenty minutes. If a suitable class-room be available, a scheme whereby it may be converted into an Industrial Room will at once suggest itself to the intelligent teacher. Further hints upon this subject are given hereafter. We feel assured that any teacher who has room and time for teaching Geography, Grammar, and Arithmetic, can profitably spare at least a small share of both for Manual instruction.

*Fourth.* In the Primary Grade the *material* first to be used will consist of the blocks. These may be purchased at small cost, or made in accordance with the directions on pages 40 and 41 of this book. For a class of fifteen pupils, will be required seventy-five cubes and fifty each of half-cubes and oblongs; or, fifteen of the sample **boxes** described on page 40, with an additional supply of half-cubes and oblongs.

For stringing straws, the material should consist of straws prepared as described on page 70. There should be at least fifteen of each length, for each pupil. For stringing beads, materials can be procured as indicated on page 80. An ounce of beads of assorted kinds will be required for each pupil.

For Learning Colors, small exertion and no expense combined with *large interest*, on the part of teacher and pupils, will procure an abundance of material. If ever at a loss for fresh illustration, it must be remembered that the book of nature is always open. Special instruction under this head will be given hereafter.

For Tablet Laying, materials may be readily secured very cheaply as indicated on page 90. The four kinds mentioned are the square, oblong, equilateral triangle, and isosceles triangle. At least fifty of each of these kinds will be required for a class of fifteen pupils.

From its nature, the material used for Paper Folding cannot be satisfactorily used but once. But it is easily procured and prepared (see page 95) and so, if properly used, will not necessarily make much expenditure.

The materials thus far mentioned, comprise those which are introduced in the first year of the Primary Grade.

The second year, stick laying, picture cutting, scrap book making, spool work, paper embroidery, and braiding are introduced. A method for procuring, preparing, and presenting the materials required for these occupations will be found described under each specific head, in its proper place in this volume. The amount of material needed must depend upon the number of pupils to receive benefits therefrom, and must be decided by the judgment of the Instructor in charge.

For the third, fourth, fifth, and sixth years of the Primary Grade, the materials necessary for an Industrial course will be found fully detailed and described, each under its proper head in the pages which follow.

Further information desired, upon the subject of Industrial material, can be gained by application to E. L. Kellogg & Co.

A complete set of materials and implements, sufficient for the introduction of Manual Training into any school, can be secured upon application to the firm above mentioned, or from parties to whom they will refer the applicant.

*Fifth.* *A few cautions* should be observed in introducing Manual Training. In the first place, too much should not be attempted. The teacher, however ambitious, will perhaps have to be satisfied with short time, few materials, small classes,—and those of the lowest grades,—and inconvenient arrangements. If he is in earnest and knows how to do the work, the pleasure he will derive from watching the growth of the enterprise will amply compensate for the discouragements of the first outlook. He must make the work popular by the power of his own interest in it. He must make it so attractive that the children will like it and like books because of its indirect influence. He must be economical; using such materials as require small expenditure of money; such as can be used and re-used. However cheaply or easily the material may have been secured he must remember that nothing will justify extravagance or wastefulness in its use.

He must use tact and wisdom in introducing, conducting, and enlarging the work. A natural appe tite is often spoiled by overfeeding. The work must be needed and asked for, and the worker stimulated to the best exertions of which he is capable, by a knowledge of something beyond, worth striving for, which honest striving will obtain.

## MANUAL WORK FOR GIRLS OF THE GRAMMAR SCHOOL

may be properly introduced as soon as the advancement of the work in the lower grades seems to warrant the step. As in the case of the Primary Departments a gradual introduction of the work is advised. Let the classes be few in number and called but two or three times during the week, and then do not allow the sitting or recitation to continue longer than forty-five minutes. Let the first work be a review of that done in the final Primary year, with just enough of variation and introduction of new occupations, to insure interest. Attempt little and be satisfied, at first, with simple work and modest results. Allow any pupil to take home with her any satisfactory piece of work which she has completed. Active co-operation of patrons cannot be expected unless they have some knowledge of the enterprise, its object, its working, and its results. In the Grammar School a class of girls for Manual work should not contain more than seven members.

THE TEACHER of Industrial work in the Grammar School must be governed by the same general conditions as those above stated in reference to the Primary Department. In this case, however, a special teacher for the Department of Manual Training is a greater necessity than in the former. This is demanded by the age of the pupils, the nature of the occupations as the work advances, and the general conditions of the Grade. If means are narrow and ways limited, still a determination to succeed will go far toward success.

PREPARATION FOR THE WORK is an absolute necessity on the part of the instructor if he would succeed. How is this preparation to be obtained? The door of the Training School does not stand open to all, but if it did, it is not the only way to knowledge of the subject and method of instruction. The wide-awake, practical teacher will learn how to teach Manual Training by reading books and educational papers, by asking questions, by experimenting, by inventing, by making good use of common sense.

THE ROOM AND TIME for Manual Training in the Grammar School may be subject to limitations at first, but successful growth will suggest and promote ways and means. It has been proven that the substitution of Manual for Mental work at stated times, say two or three times in the week, has resulted favorably for the progress of the latter. The ordinary school room may be used, but a recitation room is to be preferred if available. (For further suggestions see page 186.)

THE OCCUPATIONS for the first year in the Grammar School are those which may be classed under the general term,—plain sewing, viz.: sewing over-and-over, running, hemming, stitching, overcasting, and gathering. For this work, to supply a class of seven pupils, six or eight yards each of bleached and unbleached muslin will be sufficient material for a beginning. The supply can be readily replenished when necessary. A few yards of calico may be added if desired. Add a dozen spools of white thread, Nos. 40, 50, and 60; a half-dozen papers of good needles, No. 8; a convenient pasteboard box for each member of the class, and an additional supply for general uses; a dozen cheap thimbles of various sizes; a paper of good pins; several pairs of shears and scissors; some pieces of bees-wax and a tape-measure, and the simple outfit may be considered fairly complete. (For manner of preparing and conducting the work the teacher is referred to page 191.)

The occupations for girls in the second and third years of the Grammar School are knitting, crocheting, patching, darning, and making button holes. The quantity of yarn, worsted, and fabric required will depend upon the amount of work and size of the class. In introducing the work, it would be well perhaps to order, at one time, simply the quantity needed for the work in hand. The supply may be augmented as necessity demands. A crochet hook and set of good knitting needles will be required for each member of the class.

It must not be understood that the curriculum here laid down is exhaustive. If the teacher loses sight of the value of *original device*, the best benefit of any occupation is missed.

Ample directions for the introduction of printing and type-setting may be found under the proper head in this book.

## MANUAL TRAINING FOR BOYS IN THE GRAMMAR SCHOOL.

### SEVENTH, EIGHTH, AND NINTH GRADES.

This is a new field of labor, requiring special preparation; none, however, that may not be readily devised by the enthusiastic, determined teacher.

THE INSTRUCTOR. If the means are ample, and the spirit and purpose fully developed, a special teacher or foreman should be employed, and placed under the general direction of the principal. Otherwise the principal himself must take the work in hand assisted by one or two of the older pupils. It is understood of course that the instructor, whoever he may be, has a good knowledge, theoretical and practical, of the work to be done by his pupils. Assistant pupils, acting as foremen, should also be well posted in the work of those under their care.

THE ROOM. With money in hand, of course, a shop may be easily provided (see directions on page 199) with an outfit of tools, as they are required. A small beginning may be made by fitting up one or two knock-down benches (see page 200), and setting them up in the school room or other convenient place, at the close of school, and a class of four boys put to work with them, for thirty or forty-five minutes. This plan, as a beginning, has been employed with success. Again, a neat little cottage may be put up at small expense, on the school grounds, its size depending on the number to be accommodated. This plan has also been successfully tried, when first introducing the work, and it may be made permanent. There must be a work room of some kind, and the thoroughly in earnest, enthusiastic teacher will not fail to devise the means for providing one.

THE TIME. If the shop is kept open during school hours, the program must be made so that the principal can be at liberty fifteen or twenty minutes once or twice a day, in order to oversee the work, and the work in the shop must not interfere with the recitations of the pupils. A little attention should be given to these matters at the opening of the term. Each class should have forty-five minutes two or three times a week.

THE TOOLS. The only tools required for the first series of lessons are the hammer and the nail set. There should be light hammers for the smaller boys and full sized ones for the larger boys. The number should be governed by the number in the class for work at the same time.

For the second series of lessons, it will be necessary to have rules, squares, try-squares, thumb-gauges, straight-edge poles, knives, scratch-awls, and chalk and lines, not for each member of the class, but enough for them to use. Two or more can work together in these lessons.

For the third series of lessons a supply of saws must be had, cut-off saws, rip-saws, and back-saws. As all the members of the class will not use the same tool at the same time, economy may be used in the purchase of the tools. Two or three saw-horses will be required.

The tools required for each series of lessons are indicated at the opening of the chapter treating of each series. The cost will depend upon the quality and the number required. There are so many grades or qualities of the same tool that prices will vary greatly, but it will generally pay to get the best.

THE MATERIAL. For the first series of lessons, a few pieces of planed pine and hemlock and three sizes of nails, 4s, 6s, and 8s.

For the second series, a few planed pine boards, six or ten feet in length, and other shorter pieces of different widths.

For the third series, the lumber like that given for the second may be used.

The amount for each of the series of lessons will depend upon the number in the classes, the work done and the trials to be made before the work is accepted by the instructor.

It is believed that when the first three series of lessons have been taught, the instructor will be fully prepared to provide tools and materials for further work without any directions or suggestions.

THE INSTRUCTION. The best results will undoubtedly be achieved when the young workman studies carefully in the book each lesson before he attempts to do the work. With the book in his possession he can be fully prepared for an intelligent trial with the tools before going to the shop; he will need less attention from the instructor and a larger class can be taught at the same time. *Every piece of work completed by the pupil should be submitted to the instructor and accepted before another is taken in hand.*

REMARKS. Be careful not to undertake too many kinds of work at the beginning, and master each kind before taking up the next. Do not expect too much of the pupil; if he does the best he can it must be accepted till he can do better.

The lessons given in this book are not exhaustive, and hence the way is open for original devices, new lessons, and other methods. The instructor should not lose sight of the fact that the mind of the pupil is to be actively employed, and that the results of his efforts should receive all due encouragement and praise from both his teachers and his parents.

One side of the sewing room and shop both should be fitted up with pigeon-holes 12x15 inches and 15 inches deep, smaller or larger, in which each pupil may place her or his work when about to leave the room.

### IN THE HIGH SCHOOL,

Manual Training for girls will have an acknowledged place if its introduction in the lower grades has met with favor and resulted in success. Its introduction here will follow naturally. In the experimental stage, most of the necessities and demands of the case may be covered by the suggestions given above, in regard to the work in the Grammar School. (Further information may be gained by consulting pages 253–262.)

FOR THE BOYS IN THE HIGH SCHOOL. The work herein is a continuation of that done in the Grammar School, and under similar conditions. The tools are mostly already in the shop; and the material can be purchased as it is needed, if it is not kept in stock.

Further directions will be found in the lessons themselves.

# PART FIRST

## *THE CLAIMS OF MANUAL TRAINING.*

THERE are many questions regarding Manual Training or Industrial Education, which present themselves to the practical mind and demand an answer. Most prominent among them is:

I. **Upon what grounds is the claim made for the introduction of Manual Training into our schools?**

1. *It is believed that it ranks in importance with the study of Numbers or Language in the benefits it confers on its recipients.*

By training the eye and hand we educate; it is manual work that appeals to these in a most persistent manner.

"Nothing stimulates and quickens the intellect more than the use of mechanical tools. The boy who begins to construct things is compelled at once to begin to think, deliberate, reason, and conclude. As he proceeds he is brought into contact with powerful natural forces. If he would control, direct, and apply these forces, he must first master the laws by which they are governed; he must investigate the causes of the phenomena of matter, and it will be strange if from this he is not also led to a study of the phenomena of mind. At the very threshold of practical mechanics a thirst for wisdom is engendered, and the student is irresistibly impelled to investigate the mysteries of philosophy. Thus the training of the eye and the hand reacts upon the brain, stimulating it to excursions into the realm of scientific discovery, in search of facts to be applied in practical forms at the bench and the anvil."*

These words of an able writer, a careful thinker and student, and a gentleman of large experience in schools for manual training for boys, apply equally well to the girl, who is early initiated into the mysteries of doing and making. She too learns that the mind must impress itself upon matter through the eye and the hand, and that by her skill in so doing she may become a useful member of society, discovering truths and receiving and conferring benefits. It is to her a priceless endowment.

2. *It aims at positive usefulness.*

The belief is becoming widely spread, that any system of education that does not have for its object in some degree the self-sustaining of the recipient, that does not aid him in becoming a producer, is radically defective. Fifty or sixty years ago, an education in the schools of that day was a sufficient guaranty for a livelihood. The recipient could step into a place in business, commercial or professional life, and woo and win success. There were few competitors for the places of educated men and women, and they were the acknowledged superiors of the uneducated masses. This is not the case to-day.

An education obtained from books and competent instructors does not necessarily aid one in becoming

*C. H. Ham.

a good bread-winner. It is a noticeable fact that very many of our youth who complete the course of instruction in our high schools have but one resource for obtaining a living by their own labor, and that is teaching. They are effectually shut out from the rest of the world until they shall have learned to do something, learned to use their hands in some useful industry; and so in sheer disgust they not infrequently give their education to the dogs, and go to work; but not always. The ranks of the idlers and the demagogues are often replenished by them.

3. *It gives a feeling of independence of character to the pupil.*

Children learn to take practical views of life, and to form and maintain a close alliance with the things from which they are fed, clothed, and sheltered. The healthiest, clearest, most vigorous minds are usually found in bodies which tread the earth with the proud consciousness that they know how to appropriate and use some of the good things contained therein. The reason that self-made men and women take their places among the great and powerful, is because they were compelled, through necessity, to become familiar with work, to learn to do things, as well as to study books. They have had manual training all their lives. Wendell Phillips says: "The best education in the world is that got by struggling to get a living."

4. *Human beings were meant to employ their muscular powers.*

In the education of our youth we proceed upon the theory that they are purely intellectual beings, and hence that the pursuits of the school should be for the sole purpose of developing the intellect. Now the fact is, children, as well as men and women, are animals, having physical organisms to be trained and developed. The physical nature makes it necessary for them to know something of the physical world, objects that they can see, taste, and handle with their hands. And the better they know this outer world, how to mold, change, transform it, to a useful purpose, the more likely are they to succeed; for the fittest survive. The earlier they begin to experiment with the physical world, the better hold do they have of life, and the more serviceable may they become to themselves and others.

5. *Children love activity.*

The great majority of children and youth prefer to do something, go on errands, do some kind of work that is pleasant, agreeable, to the drudgery of study, trying to obtain useful information from a book. Call for a messenger in a school-room full of pupils and three-fourths of them will jump at the opportunity of getting away. The life of a student is not natural to them, and notwithstanding the best and most wisely directed efforts of the best instructors, they will give up efforts in that direction, when once free from the restraints of the school-room. They don't expect to live by study. They don't want to, and they will not. They do expect to work, and if they were early and wisely trained to use their powers in learning to do a great variety of things, it would help them materially in the high road to a good living, and aid their aspirations for success.

6. *Children are born to be doers rather than learners of book knowledge.*

We must remember that the brains of the young are in a formative condition, and are not a perfect medium for the action of the mind, that the food taken should to a large extent go to nourish and develop the body; that the growth and development of the brains, the mental powers, should be natural, and not in any sense forced, and that all intellectual work should be as nearly as possible voluntary and without restraint. Hence the admonition to make the duties of the school-room pleasant and attractive, in order that the pupils may be drawn rather than driven thereto.

And yet we all most thoroughly understand that idleness, listlessness at school, is pernicious, fruitful of evil, and none better than those of us who think we have discovered, that under our present system of education the young minds are liable to be overworked, that they are "fetched through" their annual courses at a sacrifice of mental vigor for future attainments, and are not infrequently made dull and stupid, instead of active and bright by persistent overwork. We know too that occupation, employment, is the foster-mother of industry, that all legitimate success in life depends upon the constant use of the powers of the body and mind. So we say with a due degree of reserve and modesty, that the Manual Arts mingled with the study of books and things, suggest a reform much needed, and which should be introduced all through our schools at no distant day.

7. *At home the child is trained in doing.*

The movements of a child, with the body, the hands, the eyes, are at first without meaning, mechanical, emotional, more or less unconscious; but by experience and training, they become intellectual, more or less under the direction of the mind. The infant in learning to talk uses words without knowing the meaning of them, and often with no definite purpose of expressing an idea or thought; the incoherent utterances are simply manifestations of a fullness of animal life, an overflow of spirits; but through the careful, skillful efforts of the mother and friends, and the influence of companions, these unmeaning expressions rise to an intellectual apprehension of their meaning.

8. *The school should continue what is begun in the home.*

The first efforts of the child at school are more or less mechanical, unmeaning; but by experience and practice the mental processes are mastered and at length become more and more complicated.

In a similar condition we find them with regard to the use of their hands and eyes, in forming, molding, and making objects useful and beautiful; and in a similar manner, if at all, must these powers, faculties, be trained and developed. So, as we believe it wise to have our children grow to manhood and womanhood with some knowledge of Language, Mathematics, and the Sciences, why not also give them some practical knowledge of the things they live by, are clothed with and sheltered under? Why not give them instruction in the Manual Arts? We do not teach the professions in our public schools, nor should we the trades, but simply give them as much instruction, experience, as will enable them to choose wisely when finally they are ready to enter upon their life work. A young man who, at this period of his life, knows his powers for work with his hands, who has been trained more or less in the Manual Arts in addition to intellectual culture, has a decided advantage over those who have not; and why not in the name of human progress give it to him in the school, as the chances are that he will get it nowhere else.

9. *Very many children dislike books.*

In all of our schools we find a greater or less number of pupils who for some sufficient cause do not take kindly to books and study. They may be faithful, but they are weak; they may be willing, but they cannot accomplish their tasks; they may be dull, but they are also indifferent. These classes of pupils cover a large field of the educational work of the teacher. And what becomes of them? Some of them fall out early by the way, some continue the unequal struggle, blindly groping their way, until thick darkness envelops them and they too fall out. Others still hold on, encountering persuasion, entreaty, threatening, driving, until at last they become incorrigible. They not only cannot, but they will not, be profited by the best efforts of the instructors. Now I undertake to say that almost every individual in these classes may be reached, may be made intelligent and appreciative, may be made to forsake evil ways, and become attentive and obedient, by simply putting his senses and his hands into harmonious relations with his mind.

Training in the Manual Arts will accomplish this very desirable result, in many cases, as I know from repeated trials; and under the most favorable circumstances will seldom fail to deeply interest the otherwise indifferent, stupid and incorrigible. The unfolding of the mind is well nigh an accomplished fact, when it is attempted through the use of the eyes and the hands. The world becomes beautiful and attractive to those who are engaged in making it so.

Now suppose that all our children and youth, male and female, were trained in the Manual Arts according to their circumstances and capacities, the same as they are now in the languages and mathematics, would not many of these harmful features of society be remedied, and the world be the better for it? Whatever others may say, or think, or do, I believe that this reform is to be the saving grace of the American republic, and the nations of the old world as well.

10. *There is a growing distrust of the methods of the public schools.*

Are not many people thinking and saying that the curricula of our primary and secondary schools need to be revised and corrected? There is a suspicion abroad that young children, especially in our graded schools, are trying to learn too much and too many things from books; that the confinement of the school-room is not evenly adjusted for the growth and strength of the body; and that so much study of subjects above their powers to comprehend, has a tendency to weaken their love of knowledge. It is coming to be believed that it is the height of unwisdom to keep young children, and youth as well, all the day at work upon subjects in which they have little interest, to appreciate which they necessarily have small powers, and weak memories to retain what they neither understand nor care for.

The difference between the study of books and learning the Manual Arts is just this: with the former, after the work of committing to memory, comes the examination, during which we undertake to determine, by questions not always legitimate, the contents of his storehouse of learning; the result is not satisfactory to either teacher or pupil, because the interest being gone, the memory fails to retain all or nearly all that is beyond his years and experience. But training in the Manual Arts being actual experiences, they remain in the mind for suitable digestion and assimilation; and there being no premature search after what he knows, no fear of catechists and catechisms, his thoughts ripen and we see and he feels that he has gained some knowledge.

11. *It is said that our schools promote laziness.*

School life is necessarily more or less sedentary. The young person who attends closely to his daily duties, having three or four studies to master, finds but little time for physical exercise, and the inclination to a quiet, contemplative life grows as time advances. Work with the hands, or any duty that requires bodily activity, becomes distasteful; and this kind of inactivity finally begets indolence of both mind and body. The habit of inaction thoroughly formed, it requires a stronger purpose than the ordinary young person possesses to overcome. Once lost, the desire to engage in those active duties by which fortune and honor are achieved seldom returns.

Under the present state of things, if a young man finds himself without a disposition to work, and he does not have to look far usually, or if a farmer or mechanic has a son with a delicate or diseased constitution, he gets what is called an education, and enters upon one of the professions, and so the ranks of statesmen, lawyers, doctors, preachers and teachers are freely and constantly replenished with weak, inefficient, and untrustworthy men.

12. *Manual Training has caused satisfaction where introduced into schools.*

The best, the strongest argument to any one is the actual test of the workings of the Manual Arts in the schools. During all the years (some eight or ten) that the subject of Manual Training has been under discussion, and the work of introducing it into the schools of Jamestown has been going forward, the writer can say that not one word has been uttered in opposition. Doubts were only expressed as to its feasibility and usefulness; and these so far as is known have yielded to the test of actual experiment. Patrons have expressed their pleasure in view of the new interest manifested by their children in the duties of school; the teachers have always been more than willing to coöperate in developing every feature of the work; and nearly every pupil that has been directly connected with the Manual Training department is a living witness of some good received. Nor is the claim unjust that the general management, the government and the instruction of the schools, have been made less laborious and more attractive by its introduction. And the consciousness of having in some degree been instrumental in putting the children and youth in possession of certain solid advantages, affords a better assurance of having done good than the usual work of a teacher's limited career.

13. *Manual Training will greatly improve our present school system.*

There is no proposition to give up the substantial features of our school system; but only to add to the course of study, or to substitute for things of less importance training in the Manual Arts. It is the belief of many thinking men and women, that our present system can be immensely improved. If it is the province of education to assist the recipient in obtaining a livelihood, if a part of its purpose should be to make the young familiar with the outer world, to teach the relations of matter and the development of force, power, practically, if its aim should be to kindle in them a desire to *do* something, as well as to study books, to become workers in the great hive of the industries; if its object should be to develop muscle as well as brain; if it is the part of wisdom to give a portion of the hours of school to establishing harmonious relations between the eye, the hand, and the mind; if it is a good thing to teach the pupil that occupation, constant and wisely directed, is the foster-mother of industry, through which comes all legitimate success in life; if the instruction of the school-room should be so directed as to encourage a close and friendly alliance between labor, brains, and capital; and if training in the Manual Arts will materially aid in accomplishing these grand objects, stimulating and quickening the intellect, who shall say that they should not be introduced into all the schools of the land? It will probably be found that here are the leading features of the new education, of which we have dreamed.

14. *Manual Training promotes human progress and happiness.*

In all the civilizations of this age, and all past ages for that matter, there has been a constant tendency to divorce labor from brains, and it has generally happened that brains was the plaintiff in the suit. Capital has always held the mastery over labor, because it has carried the brains, the requisite mental vigor and foresight, to keep in subjection a partner that ought to stand by its side, equal in honor, equal in authority. If there were 10,000 Powderlys in the United States, or men of intelligence and honesty, socialists and anarchists would be relegated to the shades whence they came and where they belong, and capital would have to share the burdens of society with greater liberality; the great problem that agitates with fearful apprehensions the thinking men and women of the land would be solved. If the workmen in the land were as intelligent, as cultured, as they ought to be and might be, there would of necessity be a healthy union of labor and capital, and the prosperity and progress of all classes would be assured.

But another question will arise:

II. Is it practicable to add Manual Training to the present course?

If it ranks in its benefits to the recipient, with the study of Numbers or Language, it should, and it will in good time take its place beside them in the curricula of our schools. So in proportion as it is seen to be useful, as we learn that it constitutes an essential element in the education of youth, it will become a part of their daily duties in the schools. Hence the test of the practicability of training in the Manual Arts in our schools must be made by determining the degree of its usefulness; and I venture to suggest, that it would be the part of wisdom if some of the recognized subjects of study could be more carefully measured by the same standard. A course of study to which Manual Training is added, the enlightened educator will find to be more practical and healthful than the usual one that demands book work only. It is just the addition the anxious parent would call for; it gives a pupil something to do. It is as practicable to teach Manual Training as to teach Language or Numbers or any other subject pursued in the schools. The additional expense is no more than has been freely accorded to the progress of civilization on every other hand. If it were added by substitution, it would assuredly be a wholesome relief from useless study and depressing confinement. The verdict, so far as it has been tried, is, "eminently practicable." And the impression is rapidly growing in the minds of the people, that it is to become one of the essentials in the education of children and youth.

During a trial of about six years, in Jamestown, there has been a constant growth of the department until now a great majority of the pupils in the schools receive the benefits of the training to some extent, and in the near future we hope to be able to offer to all a full course in the department. While our Board of Education would not have thought it possible or even wise to undertake to establish a course of Industrial Education in the schools, they have promptly and unanimously seconded every effort to give the department a permanent place in the curricula, and any effort to displace it, or any way to cripple its operations, would, I am sure, meet with their determined opposition. The practical teacher will next ask, if these things are true:—

### III. How can Manual Training be given in our schools?

1. It must be introduced without doing any injury to the cause of education, but rather serve its best interests.
2. It must be organized as a department of instruction, and managed and represented as one of the essential features of an education.
3. If it is to be given a permanent place in our public schools, it must be done eventually by authority of law; suitable legislation must be had, followed by specific direction of the Superintendent of Public Instruction, in the same manner as Drawing was introduced into the graded schools of the state some ten or twelve years since. But since immediate legislative action cannot be looked for, at least until further tests of its advantages have been made,

and the means of its accomplishment determined, voluntary efforts will have to be made by different institutions throughout the state. The press, and thinking men and women, are already giving some attention to the subject, and we may confidently hope that public opinion will ripen so that a decision may be anticipated in the not far distant future.

In the mean time, it may be placed in those schools whose teachers are desirous of solving the problem for their own satisfaction and the public good, and the greater the number of the schools that engage in the work, the sooner shall we have the required legal enactments.

MANUAL TRAINING IN THE PRIMARY SCHOOLS.—The work of Manual Training in these schools is not complicated nor expensive, and is very easily managed; hence it can be easily introduced into them. The subjects of study and practice in our public schools may be classed as follows, viz.: 1. Languages; 2. Numbers or Mathematics; 3. Objects (in the higher grades, Sciences); 4. Manual Training. In organizing a primary school for work, each class or grade should be given one or more studies under each head; thus the class will have one recitation each day in Manual Training. If time is wanting, the class may be called on alternate days. When the class in Manual Training is called, give to each pupil at his desk or on the recitation seat, on his slate, the material as provided, and set him at work. The course of study and instruction for the primary grades will be found in Part II. of this book, and the lessons showing how a class in Manual Training may be conducted will be found in Part III.

It is claimed that the pupils under this course of instruction will accomplish more, and do better work than if Manual Training were omitted.

MANUAL TRAINING IN THE SECOND PRIMARY SCHOOLS.—These schools or classes, constituting the 4th, 5th, and 6th grades, are engaged in advanced work on the same subjects as the Primary, or 1st, 2d, and 3d grades. The classes should therefore be called and conducted in the same manner. In all of these first six grades the boys and girls should be placed in the same classes in Manual Training and do the same work. The courses of study and instruction for these grades will be found in Part II. of this book, and the lessons showing how they may be conducted will be found in Part III.

MANUAL TRAINING IN THE GRAMMAR SCHOOL.—In these schools or departments, which constitute the 7th, 8th, and 9th grades, the work of organizing and conducting the classes in Manual Training becomes a little more complicated. There must be a shop for the boys, or a place in which a certain number of them can work at the same time; and a sewing room for the girls. In both rooms there must be competent instructors, other than the teachers of the class rooms. The shop and sewing rooms may be vacant rooms fitted up, or basement rooms well lighted, or a building for the purpose near the schoolhouse. If capable instructors cannot be secured, place the rooms in charge of some of the older pupils in the high school, under the general direction of a class-room teacher, giving each of them the care of one class every day. Send to these rooms as many boys and girls as can be accommodated at one time, to remain from 45 to 60 minutes. They can go every day, or on alternate days, as convenience or necessity may require. It may be needful for the pupils engaged in this department to do some of their school-room work out of school hours, which they will usually gladly undertake, rather than to be deprived of the privileges of Manual Training. The work for the pupils of the grammar schools will be found fully exemplified in Part IV. of this book.

MANUAL TRAINING IN THE HIGH SCHOOL.—The students of the High School, having their recitations at different hours of the day, may go to the shop and sewing room when not otherwise engaged. They need not go in classes. They will be put upon advanced work of the same general character as the pupils of the grammar schools. The lessons to be given them will be found exemplified in Part IV. of this book. *Printing.*—Those who are to learn to set type and do other work of a printing office, are selected from the Grammar and High Schools. They may be chosen, 1, in accordance with an expressed wish to learn; 2, from an aptitude supposed to make them good type-setters; 3, from a willingness to undertake the work of learning in addition to their other school duties. The members of the class should be given one year in the printing office unless sooner relieved for cause. The number of pupils and time per day must be regulated by circumstances. Printing is one of the great industries of our civilization, and young persons well trained in the art in connection with their education will not be so liable to become tramps in after years.

THE TIME REQUIRED FOR INSTRUCTION IN MANUAL TRAINING.—In establishing and conducting a department for Industrial or Manual Training, the idea must be abandoned, that pupils should, that they can, or that they do, devote all the hours of school to study and recitations. When they have learned how to be industrious, how to train and use their memories, and where they can have a pleasing object in being industrious, they will accomplish their tasks and have time daily for other work. Plodding and droning in the school-room may be in a great degree wiped out, when there is pleasant work for leisure hours.

INSTRUCTORS FOR MANUAL TRAINING.—In almost every corps of teachers will be found one or more persons who know many things, and who learn other things quite readily. To this class of teachers may be given the charge of the primary and second department classes. In the grammar and high schools an instructor must be employed to teach the girls, which may be done in a sewing room, or in the school-room at stated times during the week. A good janitor who is a mechanic can be employed to instruct the boys in the shop, or if he cannot do it an older and more mature pupil can be put in charge of the shop, and by employing two or more of these, the shop can be kept open for work all the day. And a young man of quick intelligence, and active body and mind, can in a short time learn to take charge of the printing office. Of course these suggestions are hints for those who may propose to begin the work in a small way.

THE EXPENSE OF MANUAL TRAINING.—That there must be additional expense in establishing a department for Manual Training in our schools goes without saying, as most of the good things in this life are achieved by an expenditure of time, labor, and money, and usually the better, the greater the cost. And it is greatly to be desired that its merits be carefully and thoroughly tested by good men, in the right kind of communities, in order that the authorities of the state may be enabled to act intelligently upon the subject.

But for the purposes of experiment, a sufficient sum may be raised by donations from those interested in educational progress and reform. In Jamestown we had a fund (called the Exhibition Fund, which was raised by public entertainments given by the schools) at our disposal, and quite enough to defray the current expenses of the department while in the experimental stage. This fund we still keep good: and whenever the Board of Education hesitate or too long delay in granting requests, we draw upon our own little treasury, and if we are right, we know that our measures will be adopted by them in time.

# PART SECOND.

## Chapter I.

### *THE INTRODUCTION OF INDUSTRIAL OR MANUAL TRAINING INTO THE JAMESTOWN PUBLIC SCHOOLS.*

THERE will be, doubtless, many interested to know the origin and progress of Industrial or Manual Training in the Jamestown schools. When the writer came to the village twenty-one years ago, to organize the Union School, he had some very positive, but rather indefinite ideas in regard to the defects in the education of children and youth. The advantages afforded by our educational system fell short of the claims made for them, and the expectations of patrons were not realized. Desiring to meet demands and expectations of the public, special departments were, at an early day, established in Vocal and Instrumental Music, Bookkeeping, Physical Culture, Normal Instruction (for teachers), and Drawing and Painting.

It required persistent effort to organize these departments, and to secure teachers qualified to fill the positions of instructors in them. And when established, it became necessary to render them intensely practical, in order to satisfy the patrons and the public that they formed a legitimate feature of public instruction. The Board of Education felt that these departments paid for themselves in the added interest and patronage of the schools, and they heartily coöperated to make them permanent. But they did not feel warranted to vote all the money needed to make them successful. The money to supply this deficiency was derived from a "fund" realized at annual exhibitions given by the pupils, under the direction of the teachers; $250 to $300 were raised in this way each year. When these departments were in full operation, it was still evident to me that something was yet lacking; that our schools did not either develop the powers of the children or fit those powers for the responsibilities and duties of real life. Our youth would have to hunt up places to fit themselves for life's duties after leaving school. I thought much of this without seeing any way out.

During the years 1867, '8 and '9, it was my good fortune frequently to meet Prof. James Johonnot, one of the leading educators in the state, well known as a writer of several valuable text books on education. I had known him well for many years. But at these meetings we became more intimate, and on many occasions discussed fully and freely both the philosophy and methods of education. In these discussions was first suggested to me the idea that Industrial education would enable the school to meet the demands of the public upon it. Entertaining this new thought awhile, I labored to develop a plan for introducing it into the schools.

But the rapid growth of the schools during these years made it quite impossible to take any measure immediately toward realizing this new idea. All of the time that could be spared from my regular duties was given to assisting the Board of Education to provide suitable rooms and instructors.*

Much thought, however, was given to the subject; ways and means were devised, with a view to adopting Manual Training, but the field was so new and untried, they were always given up before trial. Occasional conversations with the teachers found them generally more than willing to undertake to carry out any measures that might be adopted. The Board of Education usually expressed themselves in favor of the theory of Industrial Education, but as it had not been tried elsewhere, counseled the wisdom of waiting until the matter was better understood. Meanwhile, we thought of it and discussed it.

In the fall of '74, it was determined to make a beginning by opening a printing office. A press, type, and fixtures, costing $125, were purchased—money from the "fund"—and set up in an unoccupied room on the fourth floor. It was placed in charge of the commercial teacher, who, when a boy, had worked in a printing office. Two classes of boys and girls of four each were selected from the grammar and high schools to learn to set type. They were given two hours or more each week, during the school year.

The Board of Education was conducting a course of lectures, and by doing a large share of their printing, the odor of "useless," "unnecessary expense,"

---

*The schools were opened in the fall of '65, with 250 pupils in all the departments and 8 teachers. In the fall of '70, there were 1,405 pupils and 28 teachers; in '77, 1,668 pupils and 37 teachers; and in '85, 2,277 pupils and 50 teachers.

"not educational," and other unsavoriness was mostly removed from the nostrils of the public. This department has been in operation since that time, with but few interruptions. It has grown from eight to twenty pupils; instead of two hours per week, four hours are given.

For two or three years, it seemed impossible to add anything more. Accident, however, opened the closed door. One day, a boy was sent to my office as incorrigible. When he came in with the note from his teacher, seeing that he was very angry, I sent him on an errand. On his return, I told him I wanted a certain article made; showed him the drawing I had made of it. He seemed greatly pleased; so I told him to make a good copy of the drawing, making each line twice as long. When finished I said,—

"Can you do this work if I give you one hour each day from school?"

"I am not a member of the school any more," he replied, with trembling lips.

"That is bad, but I think you can return if you desire."

"I would like to return."

"Then go down and handsomely apologize for your past misconduct; make good promises for the future, and you will get your seat again, I have no doubt. Please ask your teacher to come to the office a moment." Arrangements were soon made and the work done on the second trial, quite satisfactorily. Several other cases of disobedience, etc., were referred to me and similar employment was given. Some of them did their work at home and some at the janitor's bench, in his workroom in the basement.

In this unlooked for way, a little furor was created among the teachers to have more boys set to work, good boys as well as bad, and the girls too. But what should they do? Not being able to think of anything better, the boys of one class-room were set to collecting specimens of the different kinds of domestic woods. They were told how to prepare the specimens, and after twenty-five had been presented and accepted, they were offered three cents for each additional one. Sixty-one species of woods were collected and are now in our cabinet.

The boys and girls of another room were invited to collect land and water snails. They were directed where and how to hunt for them; also how to prepare them for the cabinet. And after a certain number had been collected, they were offered three cents for each additional specimen.

In the primary schools, little things were conjured up for the pupils to do. In one they were engaged in cutting and making pen-wipers of various patterns and with suitable ornamentations; in another, cutting and making picture scrap-books.

All these things made it more evident that something must be done toward making a permanent establishment of some of the industries in the schools. But the way was so untried, the public so ready to find fault, that nothing of moment was accomplished; and I sometimes felt that it would be a blessing to be relieved of the responsibility of going further forward in this new field. My assistants and I began to see that there must be a *plan*, *a regular system* carried forward which would give constant daily, or at least weekly, employment to all the pupils. And we set ourselves to work to devise and put in operation such a system. It must be done, we knew, in such a way as not to disturb the patrons and so arouse opposition; it must not, in appearance, at least, interfere with what we are pleased to call the intellectual work of the pupils.*

So materials were obtained, as beads for stringing, papers for folding, slats for plaiting, sticks for laying, pictures for cutting and scrap-book making, paper for embroidery and the accessories to use with them, and distributed among the primary teachers. A few classes were formed in these employments, and mixed with the regular subjects of study. This gave so much satisfaction that other occupations were added from time to time. In order to assist the teachers in perfecting their knowledge of these kinds of work, a course of lectures, illustrating Kindergarten employments, was given by a very capable young lady, a graduate of one of the New York City Kindergarten schools. For the girls of the Grammar and High schools at this time, '81-2, we had already put a sewing class in operation.

One of the basement corridors was enclosed by a glass partition, and supplied with tables, chairs, and material for the work. This room was placed in charge of a young lady, who was not otherwise engaged, a part of the day. Six classes of five or six each were sent to this room, selected from the lower grades twice a week for one hour. The sewing was graded, and as soon as they could do the work of a lower grade they were promoted. Each one

---

*Instruction in free-hand drawing, and exercises in gymnastics, it is to be noted, were already in full operation in all the Primary as well as the Grammar and High schools.

admitted was entitled to remain one-half of the school year. Of course but a small number could be accommodated at one time; but it was a beginning, and one which could be enlarged as opportunity offered. And in the course of the following year, the sewing-room was kept open all the day, a part of the time with two instructors. In another corridor of the basement, a single workbench for boys had been placed and supplied with tools; and two boys were sent there at a time, one to work half of the hour and the other to watch, under the direction of the janitor, or one of the young men who understood the use of tools. In the same manner as the girls, the boys were selected because they were good scholars, or good for nothing, or any other good reason presented by the class-teacher.

For a year or more this bench was occupied the greater part of the day, each boy enjoying the privilege twice a week for about one-half the year. They thus learned the use of most of the carpenter's tools, beginning with the hammer.

An opening had now been made all through the schools, not of great moment, it is true, as to the number of pupils, or the amount of work done, but it was a beginning that needed only to be enlarged, to give employment to all. The Board of Education had been apprised from time to time of the work being done, and so far as could be judged from their words and influence, public and private, it met with their hearty approval. At all events, in the spring of '82, they subscribed liberally, and assisted greatly in other ways to raise a fund with which a shop was built, large enough to accommodate four benches and three lathes, with a loft for storing away lumber, also to supply all the tools and fixtures to put it in complete running order. This done, the shop was placed in charge of two young men, and under the general direction of the janitor (a good mechanic), one of them gave instruction to classes every school-hour of the day.

Since that time changes have been steadily made in every branch of the department, all tending to improve and enlarge the methods, increase the force of instructors, and add to the number of those receiving instruction. Three years ago, two wings were added to the High School building, and in the basement are two rooms, one of which is used for a sewing-room and printing office, and the other for the shop. These rooms are about 28 feet by 37 feet, are well lighted and pleasant, and are supplied with all needed tools, material, and instructors, and are kept open during all the school hours of the day, four days of the week. The old shop has been fitted up for a kitchen, and every Friday during each term two classes of six each receive instruction in the art of cooking.

The opinions of the people of Jamestown on the subject of Manual Training in the public schools are certainly favorable, at all events no word to the contrary has ever reached me. The Knights of Labor, a very large and respectable organization here, numbering some 1,000 or more, a short time since invited me to address them on the subject, as they desired information. The following resolution was adopted by them after listening to the address:—

*Resolved*, That we, local assembly No. 2,525, Knights of Labor, do hereby tender to Prof. S. G. Love a sincere vote of thanks for his able and interesting address on Industrial Training in the public schools delivered before our assembly on Friday evening, December 17, 1886, and we congratulate the citizens of this city upon their good fortune in having such a competent instructor in charge of their schools, which under his efficient management have attained the proud distinction of being favorably mentioned as among the best in the United States.

The Board of Education has also made a permanent record of its views in the following resolutions, which were unanimously adopted January 3, 1887:

*Resolved*, That the Board of Education of the Jamestown Union School and Collegiate Institute with pleasure records its high estimation of the able and judicious efforts of Prof. Samuel G. Love in establishing as a valuable and distinguishing feature of the school, a system of Industrial Training; and also of the prompt acquiescence in, and approval of the same, by the people.

*Resolved*, That, in the judgment of the board, the inauguration of this system has largely contributed to the reputation and progress of this institution, and to extend the field of its usefulness it heartily commends the purpose of Prof. Love to publish at an early day a text-book of Manual and Industrial Training for the use of schools.

J. H. CLARK, *President*, LEVANT L. MASON, *Secretary*. Sidney Jones, W. W. Henderson, L. B. Warner, F. A. Fuller, Jr., Chas. E. Parks.

To-day, January 19, 1887, this much can be said of the department of Manual Training in the Jamestown public schools. *All the pupils in the first six grades, about 1400 in number, are given lessons daily, or at least three or four times a week, in some kind of Manual Training. One hundred and twenty-five of the girls and sixty-five of the boys receive lessons in the sewing-room or shop at least twice or three times each week, and twenty boys and girls set type in the printing office one hour four days of the week.*

Not all the pupils in the schools receive instruction in the Manual Arts, nor perhaps do those who do receive it get all the attention that is to be desired. Two things may, however, be stated with great confidence, born, we believe, of the spirit of the age in which we live, this Nineteenth century A. D.:

1st. That the time will soon come when studies yielding little discipline and culture, and less practical service to the pupil will be laid aside, and

2d. That the hour which accomplishes this end will usher in the one which affords the privilege of Manual Training to every child and youth who desire it.     SAMUEL G. LOVE.

## Chapter II.

## COURSE OF STUDY IN THE JAMESTOWN PUBLIC SCHOOLS.

*Adopted June, 1886.*

The course of instruction as at present adopted and carried forward in the Jamestown public schools, is submitted, not as a guide for other schools of a similar character, nor because it is considered as good as could be planned; but simply to show what is being done in Manual Training and the relation of that Manual Training to the general course of study.

The entire course is inserted that teachers may see what Manual Training is deemed appropriate for each grade. It has cost much thought and experiment to determine this. Certain Manual Training is appropriate for pupils beginning Numbers, certain other work for those who have made considerable progress in Arithmetic. What this should be is a matter of experiment. The arrangement here given is imperfect in some of its features, and is not satisfactory in all respects; but still it is the best that could be made in view of the attitude of the public, and the means at hand to give Manual Training.

The first six or primary grades are as follows, viz.:

# PRIMARY DEPARTMENT.

## COURSE OF INSTRUCTION, FIRST YEAR—FIRST GRADE.

### FIRST TERM.

Language—Reading; Vocabulary of twenty name words, and read from chart; Spelling, the words learned by letter; correct errors of speech; Vocal Music.

Numbers—Count objects to ten; make combinations of two objects (sticks) to 5, the teacher putting marks on blackboard.

Object Teaching—Object Lessons on common things; right and left hand and directions.

Manual Training—Writing, on the slate and blackboards; Drawing, lines and angles on the slate; Gymnastics, free and marching; Industrial, block building, stringing straws.

### SECOND TERM.

Language—Reading, continue with vocabulary and chart, and read to 20th page First Reader; Spelling, words selected from the lessons; correct errors of speech; instruction in conversation; Vocal Music.

Numbers—Count objects to 20 and learn combinations of two numbers (sticks) to 10, the teacher putting marks on blackboard.

Object Teaching—Learn the points of the compass and locate objects in the room; Object Lessons; Physiology, oral lessons.

Manual Training—Writing on the slate and blackboard; Drawing, combinations of lines and angles on the slate; Gymnastics, free and marching; Industrial, stringing beads and learning colors.

### THIRD TERM.

Language—Reading, first half of First Reader; Spelling, from all the lessons by letter and sound; lessons in conversation; Vocal Music.

Numbers—Count to 100; review combinations and represent them by figures.

Object Teaching—Locate objects in the school-room, bound the school yard, review; Object Lessons; Physiology, oral lessons.

Manual Training—Writing on slate and blackboard, the same as first and second terms; Drawing, also the same; Gymnastics, free and marching, as the first and second terms; Industrial, tablet laying, paper folding.

## SECOND YEAR—SECOND GRADE.

### FIRST TERM.

Language—First Reader, finished; Spelling, from all lessons by letter and sound; exercises in the use of words, and lessons in conversation; Vocal Music.

Numbers—Count by twos, threes, etc.; combinations of two numbers to 20, represented by figures; review.

Object Teaching—Locate and describe (bound) objects and places in the district; review; Object Lessons; Physiology, oral lessons.

Manual Training—Writing, Tracing Book No. 1; Drawing, lines, angles, and objects on the slate; Gymnastics, free, musical, and marching; Industrial, stick laying, picture cutting.

### SECOND TERM.

Language—Reading, Second Reader to 40th page; Spelling, the same as the first term; exercises in the use of words; correct errors in speech; Vocal Music.

Numbers—Write and read numbers to 500; review the forty-five combinations of two figures, read them at sight; learn the difference of two numbers.

Object Teaching—Location and direction of object, and places near and at a distance; review; Object Lessons; Physiology, oral lessons.

Manual Training—Writing, Shorter Course No. 1; Drawing, the same as the first term; Gymnastics, the same as the first term; Industrial, making scrap books, spool work.

### THIRD TERM.

Language—Second Reader to 63d page; Spelling, the same as the first and second terms; lessons on the use of words and punctuation marks; Vocal Music.

Numbers—Review of the year's work; write and read numbers to 1000; Subtraction continued.

Object Teaching—Review location and description (boundary) of places and objects, with questions on the township (city); Object Lessons; Physiology, oral lessons.

Manual Training—Writing, Tracing Book No. 2; Drawing, lines, angles, and objects on the slate, and Inventive drawing; Gymnastics, the same as the first and second terms; Industrial, paper embroidery and braiding.

## THIRD YEAR—THIRD GRADE.

#### FIRST TERM.

Language—Second Reader to 82d page; Spelling, from all the lessons, and other words selected; learn name words and review previous work; Vocal Music.

Numbers—Add columns of four figures, combining into two as they are added; Notation, elementary.

Object Teaching—Make outline of school-room, building, yard, and locate places and objects in district; Object Lessons; Physiology, oral lessons; lessons on plants.

Manual Training—Writing, Shorter Course No. 2; Drawing, review work and Inventive; Gymnastics, free, musical, and marching; Industrial, perforated cardboard embroidery, review work.

#### SECOND TERM.

Language—Second Reader to 110th page, and selections from other books; Spelling, the same as the first term; correct errors of speech, teach action words, review; Vocal Music.

Numbers—Continue column addition of four figures; review Notation, Numeration, Subtraction.

Object Teaching—Review, outlining of school-room, building, yard, and locating places, etc., and teach definitions; Object Lessons; Physiology, oral lessons; Lessons on plants continued.

Manual Training—Writing, Tracing Book No. 3; Drawing, Inventive continued; Gymnastics, the same as the first term; Industrial, slat plaiting, review work.

#### THIRD TERM.

Language—Second Reader finished, selections from other books; Spelling, the same as the first and second terms; Write sentences, giving name and action and other words, review; Vocal Music.

Numbers—Column addition of six figures, and review all previous work.

Object Teaching—Outline map of township (village or city) and review all previous work; Physiology, review; Object Lessons; Lessons on Insects.

Manual Training—Writing, Shorter Course No. 3; Drawing, the same as the first and second terms and review; Gymnastics, the same as the first and second terms; Industrial, mat weaving, review work.

## SECOND PRIMARY DEPARTMENT.

### *FIRST YEAR—FOURTH GRADE.*

#### FIRST TERM.

Language—Reading, selections from first third of Third Reader; Spelling, from all lessons in the Text Books, and other words; Language, Construction of Sentences, correction of errors, name and action words; Composition; Declamation; Vocal Music.

Numbers—Mental Arithmetic; Written, Addition, and Subtraction; Equations.

Sciences (Objective)—Local Geography; Physiology (Hygiene); Botany, description of leaves and flowers.

Manual Training—Penmanship, Tracing Book No. 3; Drawing, Free Hand, on slate and blackboard; Gymnastics, free exercises and marching; Industrial, slat plaiting, advanced crocheting, chain stitch.

#### SECOND TERM.

Language—Reading, the same as the first term; Spelling, the same as the first term; Language, the same as the first term with quality and connecting words; Composition; Declamation; Vocal Music.

Numbers—Mental Arithmetic; Addition and Subtraction, three and four figures; Equations.

Sciences (Objective)—Local Geography, Town and County; Physiology (Hygiene); Botany, description of plants.

Manual Training—Penmanship, Tracing Book No. 4; Drawing, the same as the first term; Gymnastics, the same as the first term; Industrial, paper folding advanced, review work.

#### THIRD TERM.

Language—Reading and Spelling, the same as the first and second terms; Language, the same as the first and second terms, with how, when, and where words; Composition; Declamation; Vocal Music.

Numbers—Mental Arithmetic; Addition and Subtraction, with four and five figures; Equations.

Sciences (Objective)—Local Geography and the Primary through Definitions; Physiology (Hygiene); Botany, continued from first and second terms.

Manual Training—Penmanship, Shorter Course No. 4; Drawing, the same as first and second terms; Gymnastics, the same as the first and second terms; Industrial, perforated cardboard embroidery advanced, review work.

## SECOND YEAR—FIFTH GRADE.

### FIRST TERM.

Language—Reading, Selections from the second third of Third Reader; Spelling, from all the lessons in the Text Books and other words; Language, review, construction of sentences; Composition; Declamation; Vocal Music.

Numbers—Mental Arithmetic; Written, Review, Notation, Numeration, Addition, Subtraction, with definitions of terms used.

Sciences (Objective)—Geography, Review Local, and Primary Geography through Map of the Hemispheres; Physiology (Hygiene); Botany, the same as the first year.

Manual Training—Penmanship, Tracing Book No. 4; Drawing, Free Hand and Inventive, on slate and blackboard; Gymnastics, free exercises and marching; Industrial, sewing over and over, review work.

### SECOND TERM.

Language—Reading and Spelling the same as the first term; Language, review all former work; Composition; Declamation; Vocal Music.

Numbers—Mental Arithmetic; Multiplication Table; Multiplication, one figure; Review former work.

Sciences (Objective)—Geography, review through North America; Physiology (Hygiene); Botany, description of woods.

Manual Training—Penmanship, Shorter Course No. 5; Drawing, the same as the first term; Gymnastics, the same as the first term; Industrial, crocheting, review work.

### THIRD TERM.

Language—Reading, same as the first and second terms; Spelling, from all lessons, Word Book to page 6; Language, review, relation, and emotion words; Composition; Declamation; Vocal Music.

Numbers—Mental Arithmetic; Review, Division, with one figure.

Sciences (Objective)—Geography, review, Primary through Map of U. S.; Physiology (Hygiene); Botany, the same as the first and second terms.

Manual Training—Penmanship, Shorter Course No. 6; **Drawing**, the same as the first and second terms; Gymnastics, the same as the first and second terms; Industrial, paper folding and mounting, review work.

### THIRD YEAR—SIXTH GRADE.

#### FIRST TERM.

Language—Reading, selections from the Third Reader; Spelling, Word Book to page 11; Language, review; Punctuation; Capitals; Correct errors of speech; Composition; Declamation; Vocal Music.

Numbers—Mental Arithmetic; Written, Addition; Subtraction; Short Multiplication and Division; Definitions.

Sciences (Objective); Geography, review, Primary through the Eastern States; Physiology (Hygiene); Botany, description of common plants.

Manual Training—Penmanship, Shorter Course No. 5; Drawing, Primary Drawing Book No. 1; Gymnastics, the same as the second year; Industrial, hemming, review work.

#### SECOND TERM.

Language—Reading, the same as the first term; Spelling, Word Book to 20th page; Language, review, construction of sentences, parts of speech; Composition; Declamation; Vocal Music.

Numbers—Mental Arithmetic; Written review, Multiplication and Division with two figures.

Sciences (Objective)—Geography, review, Middle and Southern States; Physiology (Hygiene); Botany, description of common plants.

Manual Training—Penmanship, Shorter Course No. 6; Drawing, Primary Drawing Book No. 2; Gymnastics, the same as the last term; Industrial, pease work, review.

#### THIRD TERM.

Language—Reading, Selections from the Third Reader; Spelling, Word Book to 30th page; Language, review, root words, derivation words, abbreviations; Composition; Declamation; Vocal Music.

Numbers—Mental Arithmetic; Written, review, Multiplication and Division with four or more figures.

Sciences (Objective)—Geography, review, Western States and Territories; Physiology, review; Botany, review.

Manual Training—Penmanship, Shorter Course No. 6 1-2; Drawing; Gymnastics, the same as the last term; Industrial, knitting, review; paper-flower making.

## GRAMMAR DEPARTMENT.

### JUNIOR GRAMMAR CLASS—SEVENTH GRADE.

Languages—Reading; Spelling; Declamation; Composition; English Language, Review words and their offices, construction and analysis of sentences, English Grammar; Latin; German; French; Vocal Music.

Arithmetic—Mental; Written, Review fundamental rules, properties of numbers, Common Fractions.

Science (Objective)—Geography, Local, Reviewed and General; Physiology (Hygiene); American History.

Manual Training—Penmanship; Drawing, Free Hand and Industrial; Physical Culture, exercises in gymnasium; Manual Training for Boys, to draw lines and lay off distances, use of the hammer, the saw, the plane; Manual Training for Girls, Plain Sewing, Running, Gathering, Stitching, Overcasting, over and over sewing, and Hemming; Printing, Boys and Girls: 1. Learn the letters in the lower case. 2. Also in the upper case. 3. To hold and handle the stick. 4. To set up and distribute words. 5. Also sentences. 6. To set up and distribute copy.

### MIDDLE GRAMMAR CLASS—EIGHTH GRADE.

Languages—Reading; Spelling; Declamation; Composition; English Language, review words and their offices, construction and analysis of sentences; English Grammar; Latin; German; French; Vocal Music.

Arithmetic—Mental; Written, Review, Decimals, U. S. Money, Reduction of Compound Numbers.

Sciences (Objective)—Geography, Local and General; Physiology (Hygiene); Botany, leaves and flowers, and general classification of plants.

Manual Training—Penmanship; Drawing, Free Hand and Industrial; Physical Culture, exercises in the gymnasium; Manual

Training for Boys, Review the work of the last year; Lessons in construction, Boring, Chiseling; Manual Training for Girls, Crocheting, Knitting begun; Printing for Boys and Girls. 7. Learn to correct proof. 8. To set up copy and distribute it on time. 9. To make up and lock forms.

### *SENIOR GRAMMAR CLASS—NINTH GRADE.*

Languages—Reading; Spelling; Declamation; Composition: English Language, review words and their offices, construction and analysis of sentences; English Grammar; Latin; German; French; Vocal Music.

Arithmetic—Mental; Written, Review, Denominate Fractions, Percentage, Profit and Loss, Interest, Book-keeping.

Sciences (Objective)—Geography; Physiology (Hygiene); Botany, description and classification of plants; Physics (Elementary).

Manual Training—Penmanship; Drawing; Physical Culture, exercises in the gymnasium; Manual Training for Boys, Review lessons of the last year, Lessons in Mitering, Doweling, Dove-tailing, begin Drawing and Construction; Manual Training for Girls, Knitting Advanced, Mending, Patching, Darning, Making Button-holes; Printing, Boys and Girls. 10. Run the press. 11. Wash type and distribute forms. 11. Do job work given out.

## ACADEMIC DEPARTMENT.

### *TENTH, ELEVENTH, TWELFTH AND THIRTEENTH GRADES.*

As the courses of Study and Instruction differ more or less in different High Schools, it is not deemed necessary that the Jamestown course of study be given here, but simply the work in Manual Training which is followed. This may be adapted to all High Schools.

FOR YOUNG MEN.
1. Drawing and Construction.
2. The Lathe.
3. Finishing.
4. Printing.

FOR YOUNG WOMEN.
1. Cutting.
2. The Use of the Sewing-machine.
3. Embroidery.
4. Cooking.
5. Printing.

# PART THIRD.

## Chapter I.

## SUGGESTIONS, LESSONS, AND METHODS OF INSTRUCTION IN MANUAL TRAINING.

### THE PRIMARY SCHOOL—FIRST GRADE.

1. THE PUPILS. These are children entering school for the first time. They are usually five or six years of age.

2. THE LENGTH OF SESSIONS AND AMOUNT OF WORK. In making a program for work under this course, it is not considered an essential feature that all the classes receive manual training every day. One subject may alternate with another in certain cases, with good results. In fact, it is better for the classes to be called at such intervals as will insure interest on their part, and a love for the exercises, rather than so frequently as to endanger their relish and appreciation of the work they are expected to perform. A longer time than fifteen or twenty minutes should not be occupied by any single lesson.

The lessons, whose general details will be given, cover substantially the necessary ground, yet variations must be originated by the teacher to give added interest and value to the lessons. New devices and plans must be thought out to encourage and help the pupils.

The amount of work planned for any distinct lesson may be divided or increased to suit the time and requirements of any program.

3. THE STUDIES AND OCCUPATIONS. What is deemed appropriate for this grade will be seen in outline by referring to Appendix, Chapter II. They cover four subjects, viz.: *Language, Numbers, Objects, Manual Arts.*

4. THE MANUAL ARTS. The occupations presented are as follows:—

>Block Building.
>Straw Stringing.
>Stringing Beads.
>Learning Colors.
>Tablet Laying.
>Paper Folding.
>Writing on Slate and Blackboard.
>Drawing.
>Gymnastics.
>Reviews.

## *BLOCK BUILDING.*

THE MATERIALS. A full set of blocks for each pupil should consist of five one-inch cubes, two half-cubes made by cutting a one-inch diagonally, and two blocks one-half inch by one inch by two inches, made by cutting a block one inch square and two inches long, lengthwise through the middle.

This set will consist of nine pieces, and can be put into a box two inches square, as a sample box. Of course, these numbers can be varied according to the work the teacher prepares to do. For example, as the pupils advance in their work, it may be thought best by the teacher to use more than the given number of any one kind, which can be taken from the general supply, or stock on hand. Beginning with one, more can be added as long as the teacher sees fit to continue Block Building.

These blocks are not expensive. They can be made easily by any person accustomed to the use of the saw and the plane, or purchased at an expense of 30 cents to 40 cents per hundred. Take a board of cherry, or maple, or any kind of hard wood, one inch thick, and two or three feet long, and cut it into inch strips with a saw. These strips may then be cut into blocks as required. The blocks should then be smoothed off, or polished by rubbing with sand-paper. Take a sheet of fine sand-paper, and tack it to a board with a smooth surface, and to this surface apply the blocks, rubbing them until quite smooth, which can be done very rapidly.

Small paper boxes can be obtained or readily made for keeping a set of the blocks for each pupil. These boxes can be made in this way: Take a piece of manila or other strong paper, six inches square; make a two-inch square in the center, and from each corner of this square cut the paper to the outside corner, then fold and crease and fasten with mucilage or paste. The boxes can be made of any size, larger or smaller than those named above. Covers can be made for the boxes in the same way, only a little larger, and not as deep.

### FIRST LESSON.

THE BOX. The training of the eye and hand is begun by the introduction of such occupations as will develop, in the best manner, correct ideas of form, size, and distance. By reason of its adaptability to produce such development, the attention and effort of the young learner are directed first to the work of Block Building.

In conducting the lesson, answers to questions may be given by the class in concert, or by individual members, at the discretion of the teacher.

The initiatory lesson, simply a study or observation of the object, is presented as follows: The teacher, holding up before the class one of the sample boxes of blocks (Fig. 1, page 53), says, "What have I in my hand?"

The pupils answer, "It is a box."

*Teacher.* What shape is it?

*Pupil.* It is square.

*T.* Of what is this box made?

*P.* It is made of wood (or pasteboard).

*T.* Since this box is square and is made of wood, what kind of a box shall we call it?

*P.* A square, wooden box.

*T.* What do we call this part of the box (indicating top or lid)?

*P.* It is called the top of the box.

*T.* How many tops has the box?

*P.* It has one top.

*T.* What is this part of the box called (indicating bottom)?

*P.* It is called the bottom of the box.

*T.* How many bottoms has the box?

*P.* It has one.

*T.* What name do we give to this part (indicating a side of the box)?
*P.* The side of the box.
*T.* How many sides has it?
*P.* It has four sides.
*T.* Count the sides (indicating as children count).
*P.* One, two, three, four.
*T.* Let us play (or suppose) for a little while, that the top of the box is one of its sides, and that the bottom of the box is also one of its sides. How many sides, in all, will the box then have?
*P.* It will have six sides.
*T.* Count them (indicating as children count).
*P.* One, two, three, four, five, six.
*T.* What name may we give to that side which is, or should be, at the top of the box?
*P.* The upper side.
*T.* Then, what may we call this part (indicating bottom of box)?
*P.* The lower side.
*T.* Of what use is the lower side, or bottom of the box?
*P.* It is the part on which the box stands or rests.
*T.* What is the use of the upper side, or top of the box?
*P.* It is the cover of the box.
  Or, It shuts up the box.
  Or, It keeps things in the box, etc.
*T.* What can we do with the cover, or upper side of the box?
*P.* We can open and shut it.
*T.* What is this part of the box called (indicating whole outer surface)?
*P.* The outside of the box.

*T.* And this part (opening cover and indicating)?
*P.* The inside of the box.
*T.* What use can we make of a box?
*P.* We can keep things in it.

The children need not be encouraged in their natural desire to tell what they know of the various uses to which boxes can be put, since that does not bear directly upon the point of the present lesson.

---

### SECOND LESSON.

THE CUBE. The teacher selects from the sample box a cube, one inch square (Fig. 2, page 53), and holds it up for the inspection of the class, saying, "What do I hold in my hand?"

The pupils answer, "A block."

*Teacher.* Of what is this block made?
*Pupil.* It is made of wood.
*T.* What is its shape?
*P.* It is square.
*T.* It is square, and is made of wood. Then what kind of a block is it?
*P.* It is a square, wooden block.
*T.* Can you name anything else that has the same shape as this block?

An immediate readiness will be manifested on the part of the children to mention objects of cubical shape which they have seen, as,

"My box is that shape."
"A tea-chest is something like it."
"A bee-hive has that shape."
"My pigeon-house is square like that." etc.

Some of the answers will call forth eager discussion, which will often result in modifications or changes, if a child discovers that he has made a poor comparison.

The teacher may now present to the class the general supply box, containing blocks of assorted sizes and shapes, and allow each child to select for himself a block of the same shape as the one which has been the subject of the lesson. Mistakes doubtless will be made in selecting the blocks, but they will rectify themselves naturally, in the course of the lesson, which will contain a double advantage for the child who has been able to discover and correct his own error.

When each pupil has been provided with a block, the teacher, placing a cube upon the outspread palm of his left hand, says to the class:—

"Each place your block, as I have placed mine."

This being done, the lesson proceeds in the same manner as the one above described.

*Teacher.* What shall we call this part of the blocks (indicating the face lying uppermost)?

*Pupil.* The top.

*T.* What shall we call this part (indicating opposite surface)?

*P.* The bottom.

*T.* What part is this (pointing to any face not before indicated)?

*P.* It is the side.

*T.* How many sides has the block?

*P.* It has four sides.

*T.* Count them.

*P.* One, two, three, four (all indicating as they count).

*T.* Turn your blocks as I turn mine (changing the position of the cube so that it rests, in turn, upon its various faces). Now tell me which is the top.

The pupils will point, doubtless, to the face which last lies uppermost.

*T.* What difference can you see between the top, bottom, and sides of your block?
*P.* There is no difference.
    They are all just alike.
    They are all tops, when they are on top, etc.
*T.* Is the bottom different?
*P.* No, the bottoms are all alike.
*T.* Then how can we tell which is the top of the block?
*P.* There is no real top.
*T.* You have found that the block has no real top or bottom, and that its sides are all alike. I will give you a name for them. We will call them all *faces* of the block. What are all of the sides of the block called?
*P.* They are called faces of the block.
*T.* How many faces has this block?
*P.* It has six faces.
*T.* Count them.
*P.* One, two, three, four, five, six (all indicating as they count).
*T.* What part of the block is formed by its faces (indicating entire surface)?
*P.* The outside of the block.
*T.* If I should bore a hole into the block, what would I find inside of it?
*T.* The same that is on the outside.

*T.* We call a block *solid*, which is the same on the inside as on the outside and all the way through. Tell me what kind of a block it is.

*P.* It is a solid block.

*T.* We call all of the faces taken together the outside of the block. We also call it the *surface* of the block. Tell me by what other name the outside of the block is called?

*P.* It is called the surface of the block.

At the close of the lesson, all of the blocks should be quietly replaced in the proper box.

### THIRD LESSON.

THE CUBE (*continued*). Each of the pupils having selected a cube as in the previous lesson, the subject is resumed, prefaced by a brief review of former work.

*Teacher.* What name may we give to this part of the surface where two faces join or meet (indicating)?

*Pupil.* It is an edge.

*T.* Count the edges.

*P.* (Indicating as they count) One, two, three . . . . . . . . twelve.

*T.* How many edges meet at this place (indicating a corner)?

*P.* Three edges.

*T.* Count them.

*P.* (Indicating) One, two, three.

*T.* What shall we call this place where several edges meet or join?

*P.* It is called a corner.

*T.* Count the corners of your block.
*P.* (Indicating) One, two . . . . . eight.
*T.* Point to a face of your block. To an edge. To a corner.
*T.* How many faces has your block?
*P.* It has six faces.
*T.* How many edges has it?
*P.* It has twelve edges.
*T.* How many corners has it?
*P.* It has eight corners.
*T.* What name is given to the outside of the block?
*P.* It is called the surface of the block.
*T.* This block has a name. It is called a cube. All solid blocks that have this shape are called by the same name.
*T.* What is the name of the block about which we have learned to-day?
*P.* It is called a cube.

This lesson is concluded pleasantly by asking the question, "How long are the edges of your cube?" The accuracy of the various estimates is tested by taking true measurements of the dimensions of the block.

The blocks are again returned to their proper place in a neat and well-ordered box.

---

### FOURTH LESSON.

MANIPULATIONS WITH A SINGLE CUBE. This lesson is opened by briefly reviewing the preceding one. Each pupil being provided with a cube as before, the teacher says, "What is the name of this block?"

*Pupil.* It is called a cube.

*Teacher.* Place your cube before you, upon the desk, so that the face nearest you will lie in the same direction as the edge of the desk (or table).

All the cubes having been placed in correct position, the pupils, following the direction of the teacher, proceed to execute the manipulations described below. Perfectly prompt attention and action should be required. The directions, as a rule, should be given but once, and as little assistance rendered as possible.

*Teacher.* Point to the upper right hand corner of the face which is nearest you. Point to the lower right hand corner. To the upper left hand corner. To the lower left hand corner.

Turn your cube one-quarter over, *from* you. Turn it half over, from you. Three-quarters over. Turn it once completely over.

Turn your cube one-quarter over, *toward* you. Half over, toward you. Three-quarters over. Once completely over.

Turn your cube one-quarter over, *toward the right*. Half over. Three-quarters over. Once completely over.

Turn your cube one-quarter over, *toward the left*. Half over. Three-quarters over. Once completely over.

Turn your cube one-quarter *around*, toward the right. Half around. Three-quarters around. Entirely around.

Turn your cube one-quarter *around*, toward the left. Half around. Three-quarters around. Entirely around once.

Point to the upper edge of the face which lies toward you. To the lower edge. To the right hand edge. To the left hand edge.

Place your cube so that an upright *edge* will lie toward you. Point to the upper corner, toward you. To the upper corner, from you. To the lower corner, toward you. To the lower corner, from you. To the upper corner, at the right. To the upper corner, at the left. To the lower corner, at the right. To the lower corner, at the left.

Turn the cube half over, from you. Turn it completely over, from you.

Turn your cube half over, toward you. Completely over, toward you.

Turn the cube one-quarter around toward the right. Half around. Three-quarters around. Completely around.

Turn the cube one-quarter around, toward the left. Half around. Three-quarters around. Completely around.

Collect and replace material as at the close of former lessons.

### FIFTH LESSON.

FORMS REQUIRING TWO CUBES. A single cube is presented, first, to each child, and the previous lesson reviewed. Then another cube is given to each one, and two blocks are manipulated by the pupils, closely and promptly following the dictation of the instructor.

*Teacher.* Place both of the cubes before you, with the faces nearest you lying in the same direction as the edge of the desk. Place their inner faces close together. What have you made? (Fig. 3, page 53.)

*Pupil.* A short wall.

*T.* Place a face of the right hand cube against an edge of the left hand cube. What have you now made? (Fig. 4.)

*P.* A broken wall.

*T.* Mend the wall again, that is, place the inner faces close together. Now move the blocks apart, the width of one cube. (Fig. 5.) What have you now?

*P.* A small, open gate.

*T.* Move them apart, the width of two cubes. What have you?

*P.* A large, open gate.

*T.* Place an upright edge of the right hand cube against an upright edge of the left hand cube. What may we call this? (Fig. 6.)

*P.* A closed gate, or, a section of zigzag fence.

*T.* Make the little wall again. Now turn both of the cubes as they stand thus together, one-quarter around toward the right. In this position make as before the broken wall. Make the small gate. The wide gate. The closed gate.

These operations may be repeated to any length desired, from different standpoints, as in the example just given.

*T.* Place one of the cubes upon the other, with the two faces which meet lying exactly together. What have you made? (Fig. 7.)

*P.* A square post, or, a little chimney.

*T.* Turn the upper cube to the right until its corners lie above the upright sides of the lower one. (Fig. 8.) What have you?

*P.* A rough chimney.

*T.* Make a smooth chimney again. Now move the upper cube upon the lower one, nearly one-half of its width toward the right. (Fig. 9.) What have you?

*P.* A little step.

*T.* Repeat the last mentioned form, moving the upper cube toward the left. From you. Toward you.

Try to build something with your cubes different from any of the forms yet made. What do you name the form that you have made?

A short time given for inventive building closes the lesson pleasantly.

### SIXTH LESSON.

FORMS REQUIRING THREE CUBES.—As the number of blocks is increased, the necessity becomes apparent, for a level surface upon which to construct the various forms. Since the ordinary desk does not afford such a surface, and tables are not usually available for school-room use, a level surface for block building may be procured in the following simple manner:—

Let each pupil place his slate before him, upon the desk, and, by inserting a book beneath the lower edge, raise the whole surface to a common level, for operation.

Forms requiring the use of three cubes are now to be considered. Each pupil being provided with this number, the lesson proceeds in similar manner to the former ones.

## TWO CUBES. 53

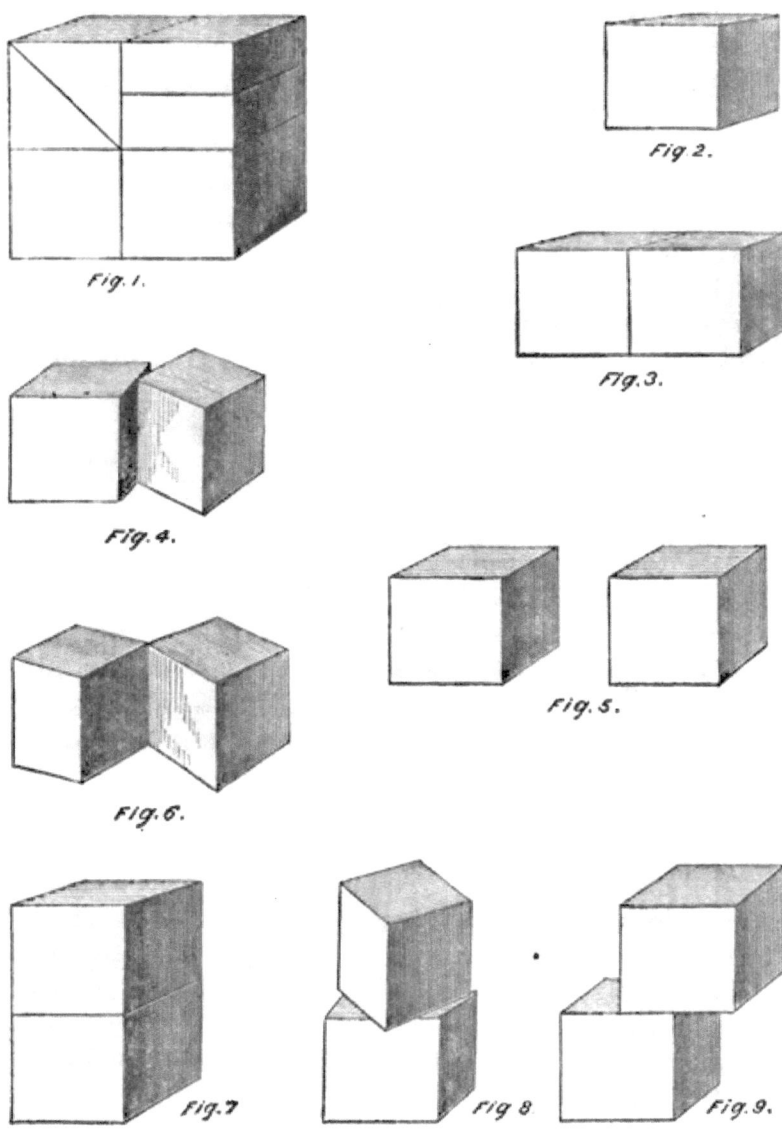

Fig. 1.
Fig. 2.
Fig. 3.
Fig. 4.
Fig. 5.
Fig. 6.
Fig. 7.
Fig 8.
Fig. 9.

*Teacher.* Place your cubes side by side, close together, each having a face toward you.

Count your cubes, from right to left, pointing as you count.

*Pupil.* One, two, three (indicating).

*T.* Count your cubes, from left to right, pointing.

Pupils count and indicate as before.

*T.* What name shall we give to the form we have made?

*P.* A garden-wall. (Fig. 10, page 56.)

*T.* Remove the middle block from the wall. What have you now?

*P.* A small gate.

*T.* Move the two cubes, a very little, toward each other. Now place the third cube above them, resting two edges upon the other cubes. What name shall we give to this?

*P.* A little covered gate-way. (Fig. 11.)

*T.* Remove the block from the top. Place it again between the other two. Push it carefully from you, until its front edges touch the back upright edges of the other two. What have you made?

*P.* A fire-place. (Fig. 12.)

A wagon, with two horses.

A safe place for a little boat to land.

*T.* Place the third cube again between the other two. Move it toward you, as you before moved it from you. What have you? (Let the pupil answer.)

*T.* Make the garden-wall again. Now move the cubes apart until the space between them is equal to the width of one cube. What shall we call this?

*P.* A row of bee-hives.

A row of bathing-houses.

Three soldiers marching along.

*T.* Place your three cubes, one upon the other, their upright faces lying in a straight line. What have you made?

*P.* A chimney. (Fig. 13, page 56.)

A post.

*T.* Take off the topmost cube. Place it at the right of the lowest one, the upright sides joining. What shall we name this form?

*P.* A little chair. (Fig. 14.)

A carriage-block.

*T.* Take off the upper cube. Place it in front of the left hand cube, upright sides joining. What have you?

*P.* A cozy corner in the wall for pussy. (Fig. 15.)

*T.* Place the three cubes in a line before you, upright edges adjoining. What is this form?

*P.* A zigzag fence. (Fig. 16, page 58.)

*T.* Remove the middle cube. Move the other two cubes a little nearer together. Place the third cube carefully above the other two, corner to corner, upright edges adjoining. What shall we name this?

*P.* A zigzag gate. (Fig. 17.)

*T.* Build again a straight wall. Remove the middle cube and place it upon the left hand one, edges nicely adjoining. What is this form?

*P.* A high stool and a low one.

*T.* Join the three cubes by upright edges only. What have you?

*P.* A little well. (Fig. 18.)

*T.* Build a small chimney, using two cubes. Place the third upon them, its upright edges above upright faces. What is this?

*P.* A dove-cote upon a post.

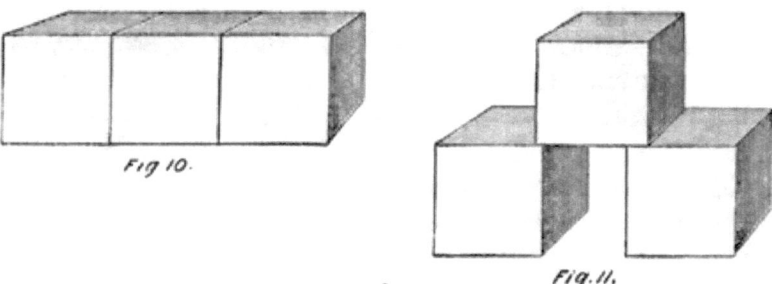

Fig. 10.

Fig. 11.

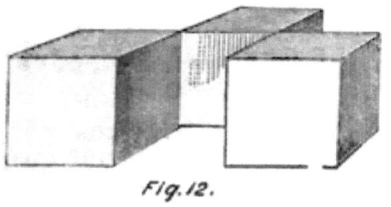

Fig. 12.

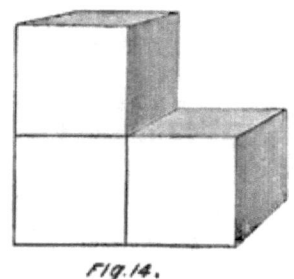

Fig. 14.

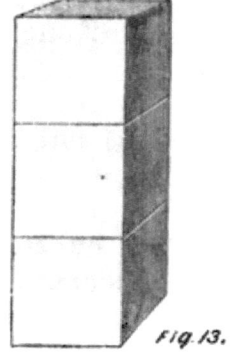

Fig. 13.

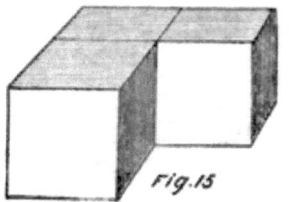

Fig. 15

*T.* Make a short wall, using two cubes. Place the third upon the other two, edges adjoining, exactly midway from either end. What name can we give to this?

*P.* A carriage-block. (Fig. 19.)

*T.* Make a wall, using three cubes. Remove the two outer cubes, and place both carefully, edges adjoining, above the middle one. What is this form?

*P.* It is a flower-stand. (Fig. 20.)

The forms described above may be varied in several ways. They may be built from any desired standpoint, that is, the position, as a whole, may be varied. The changes that may be rung, in nearly every form produced, upon the four directions, *left hand*, *right hand*, *from you*, and *toward you*, cover a large field of work most excellent for practice. Constant drill in constructing simple forms will make the little hands skillful when more complicated work is presented.

Careless or slovenly work should be regarded with great disfavor, and much emphasis placed upon the point of exact and perfect construction. The instructor should use the greatest effort to make the lessons pleasant to the pupils. All lessons in Block Building should be closed by a few minutes' work in invention. Frequent reviews of former work should not be neglected.

### SEVENTH LESSON.

FORMS REQUIRING FOUR CUBES. Four cubes are now given to each pupil, and the new lesson proceeds upon the same plan as the previous ones. After a brief review exercise, the teacher says, "Build

58  INDUSTRIAL EDUCATION.

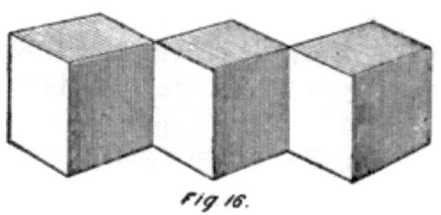
Fig. 16.

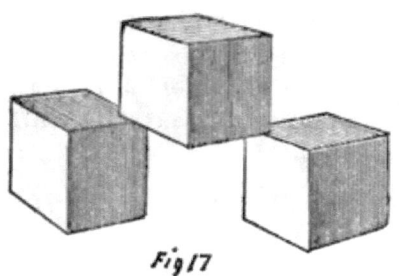
Fig. 17.

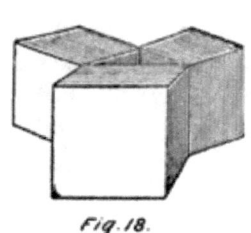
Fig. 18.

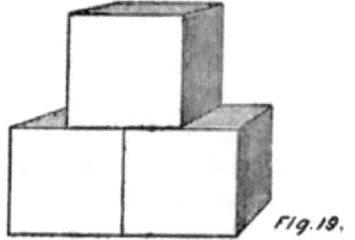
Fig. 19.

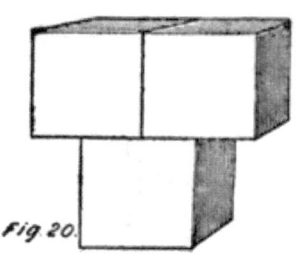
Fig. 20.

with four cubes, a straight wall. Tell me how this wall is unlike the other walls which you have built."

*Pupil.* It is longer than the others.

*Teacher.* Count your cubes, from the right toward the left, pointing as you count.

*P.* (Indicating) One, two, three, four.

*T.* Count your cubes from the left toward the right, pointing.

Pupils count and indicate as before.

*T.* Take away the left hand cube from your wall, and place it upon the next cube at the right, edges nicely adjoining. What have you now made?

*P.* A little bed. (Fig. 21, page 61.)

*T.* Remove the cube from the right hand of your row. Place it upon the upright row at the left, edges adjoining. What does this look like?

*P.* An old-fashioned chair. (Fig. 22.)

*T.* Remove the cube from the right and place it upon the top of the upright row, edges adjoining. What is this?

*P.* A chimney.

*T.* How is this chimney unlike the others that you have built?

*P.* It is taller than the other chimneys.

*T.* Take the cubes from this chimney, one by one, counting as you remove them.

Pupils remove the cubes, counting.

*T.* Build the chimney again, counting the cubes as you replace them, one by one.

Pupils rebuild the chimney, counting.

*T.* Make a zigzag fence, using four cubes. Make a row of four bee-hives. Make a short wall, using two cubes. Now make another short wall, using two cubes. Place one wall upon the other, edges adjoining. What have you made?

*P.* Another little wall. (Fig. 23.)

*T.* How is this wall unlike the other walls that you have built?

*P.* It is higher than the others.

*T.* Separate this wall into two upright rows, with space between, the width of one cube. What have you made?

*P.* Two gate-posts.

*T.* How are these posts different from the others that you have made?

*P.* These posts are taller than the others were.

*T.* Remove the upper cube from the right hand post. Place it at the left of the other post, with space between, the width of one cube. What have you now?

*P.* A high stool and two low ones.

*T.* Build again a little straight wall, two cubes high, and two cubes long. Now move the two upper cubes upon the two lower ones, nearly half their width, *from* you. What have you now made?

*P.* A small flight of steps. (Fig. 24.)

*T.* Build a little fire-place, using three cubes. Close the fire-place in front, with the fourth cube, upright edges adjoining. What name shall we give to this form?

*P.* A little well. (Fig. 26.)

*T.* Place three cubes side by side. Remove the middle cube. Move the remaining two a little nearer together. Place the other two cubes above the space thus made, edges adjoining. What have you now?

*P.* A little tower. (Fig. 25.)

## FOUR CUBES.

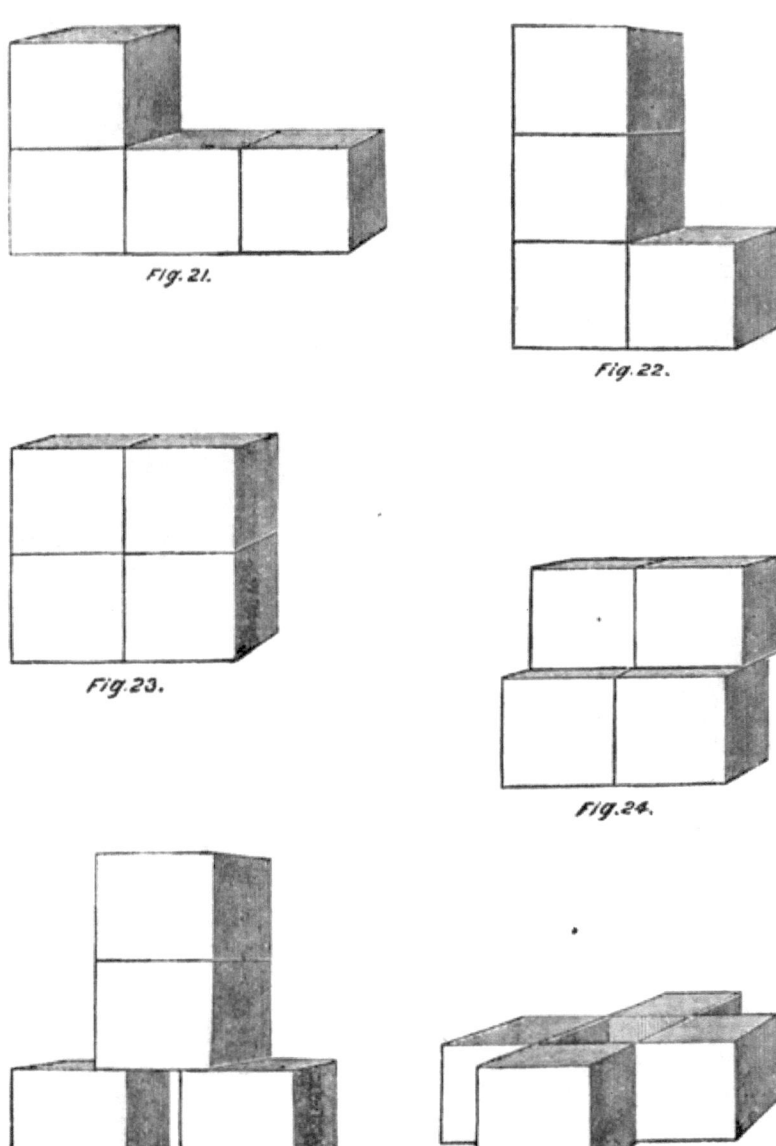

Fig. 21.

Fig. 22.

Fig. 23.

Fig. 24.

Fig. 25.

Fig. 26.

*T.* Move the two cubes at the bottom closely together. What is this?

*P.* A monument and its base.

The teacher may continue this lesson, by supplementing original designs, to any desired length. Only so much work should be presented, at any lesson, as the pupils are able to perform well.

In succeeding lessons, an additional cube may be given to each child, if desired, and work upon this single form of block may continue indefinitely. Every lesson should close with inventive work.

### EIGHTH LESSON.

THE HALF CUBE. The teacher selects from the sample box a cube divided into halves diagonally, holds it up before the class, and asks, "What do I hold in my hand?"

The answer is "A cube." (Fig. 27, page 65.)

Separating the cube and holding a half-cube in each hand, the teacher asks, "What have I now?"

The pupils answer, "A cube in two parts." (Fig. 28.)

*Teacher.* How many half-cubes make a whole cube?

*Pupil.* Two half-cubes make a whole cube.

*T.* Count them.

*P.* One, two.

*T.* (Placing the two parts together again) What have I made of those half-cubes?

*P.* A whole cube.

This object lesson upon the half-cube may be continued upon the plan given for introductory lessons upon the cube, dividing the work into convenient lessons.

Then, in successive lessons, in the same manner as that described with the cube, the pupils select from the general supply box, one, two, three, four, and finally any desired number of half-cubes, for manipulation and building.

Next in order, should follow a number of lessons employing the cube and half-cube in combination. The number of blocks of each kind used should be gradually increased, as in the lessons employing a single kind of block. For these lessons a systematic plan, such as the following, is suggested:—

    One cube and one half-cube.
    One cube and two half-cubes.
    One cube and three half-cubes.
    Two cubes and one half-cube.
    Two cubes and two half-cubes.
    Two cubes and three half-cubes.
    Three cubes and one half-cube.
    Three cubes and two half-cubes.
    Three cubes and three half-cubes.

As much of the ground above laid out may be covered at a single sitting or recitation as the instructor in charge may deem advisable, or as time and circumstances will permit. No exercise should be continued after the pupil has evinced any sign of weariness.

The teacher should aim constantly to acquire variety of method in presenting lessons to the class, especially if the lessons themselves bear close similarity to one another. Originality, both in teacher and pupil, is productive, as a rule, of the best results, whatever the work in which they are engaged.

## NINTH LESSON.

THE OBLONG BLOCK. As in the lesson just considered, a general plan for any number of exercises will again be discussed under one head.

The first lesson on the Oblong Block should follow, in all essential particulars, the object lesson on the cube. Material, shape, size, faces, edges and corners are discussed and described. Objects resembling the oblong block in shape are named by the pupils, and points of resemblance or of difference stated. The counting exercises are continued. The name oblong block, or oblong-shaped block, is given to the class.

A second lesson, comprising various movements and positions of a single oblong block, is given as in the case of the cube. (Fig. 36, page 67.)

Then two oblong blocks are given to each, then three, increasing as before the number of blocks as the class advances. All lessons are guided by the dictation of the teacher, except those in which invention is employed. Dictation should be distinct and specific, and action on the part of the pupil, prompt and accurate.

When the resources of the oblong block have been sufficiently exhausted, combinations with the cube and half-cube may be made. A great number of interesting forms may be produced from the various combinations laid down in a plan for a graded series of lessons, such as the following:—

> One cube and one oblong.
> One cube and two oblongs.
> One cube and three oblongs.
> Two cubes and one oblong.

## THE HALF-CUBE. 65

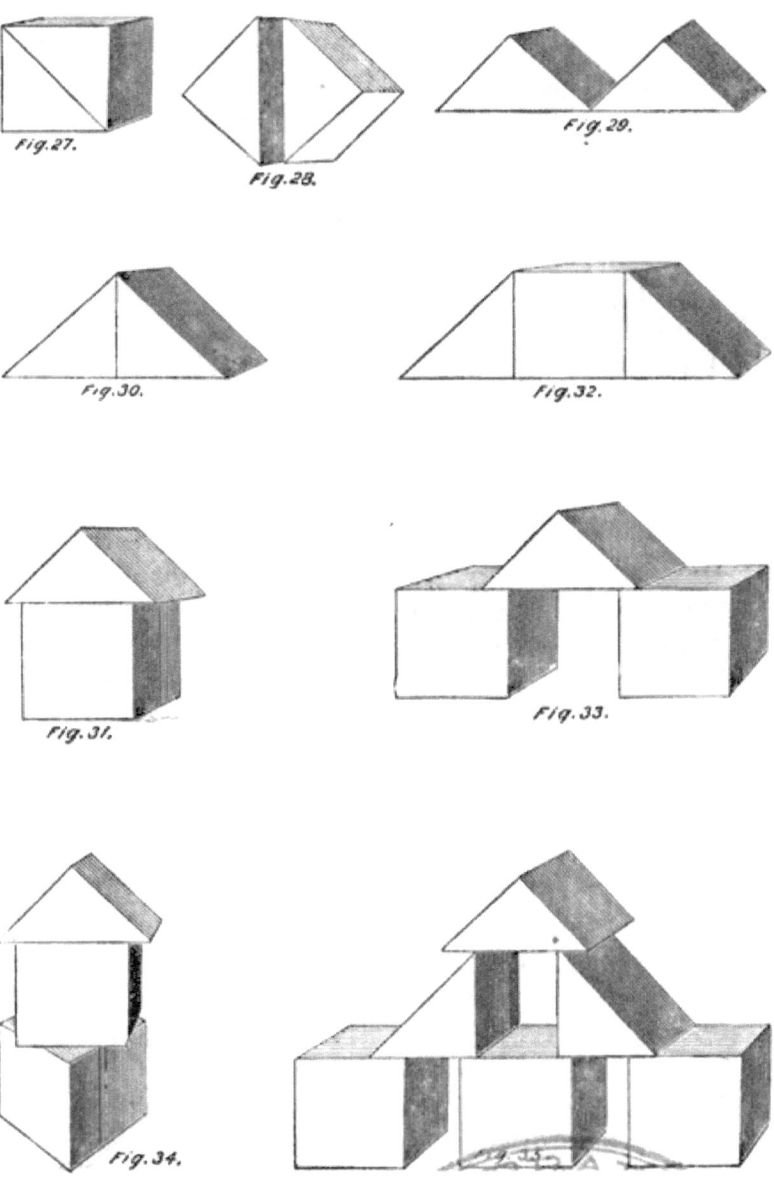

Fig. 27.
Fig. 28.
Fig. 29.
Fig. 30.
Fig. 32.
Fig. 31.
Fig. 33.
Fig. 34.

Two cubes and two oblongs.
Two cubes and three oblongs.
Three cubes and one oblong.
Three cubes and two oblongs.
Three cubes and three oblongs.
One half-cube and one oblong.
One half-cube and two oblongs.
One half-cube and three oblongs.
Two half-cubes and one oblong.
Two half-cubes and two oblongs.
Two half-cubes and three oblongs.
Three half-cubes and one oblong.
Three half-cubes and two oblongs.
Three half-cubes and three oblongs.
One cube, one oblong, and one half-cube.
One cube, two oblongs, and one half-cube.
One cube, three oblongs, and one half-cube.
Two cubes, one oblong, and one half-cube.
Two cubes, two oblongs, and one half-cube.
Two cubes, three oblongs, and one half-cube.
Three cubes, one oblong, and one half-cube.
Three cubes, two oblongs, and one half-cube.
Three cubes, three oblongs, and one half-cube.
One cube, one oblong, and two half-cubes.
One cube, two oblongs, and two half-cubes.
One cube, three oblongs, and two half-cubes.
Two cubes, one oblong, and two half-cubes.
Two cubes, two oblongs, and two half-cubes.
Two cubes, three oblongs, and two half-cubes.
Three cubes, one oblong, and two half-cubes.
Three cubes, two oblongs, and two half-cubes.
Three cubes, three oblongs, and two half-cubes.
One cube, one oblong, and three half-cubes.
One cube, two oblongs, and three half-cubes.

## THE OBLONG BLOCK. 67

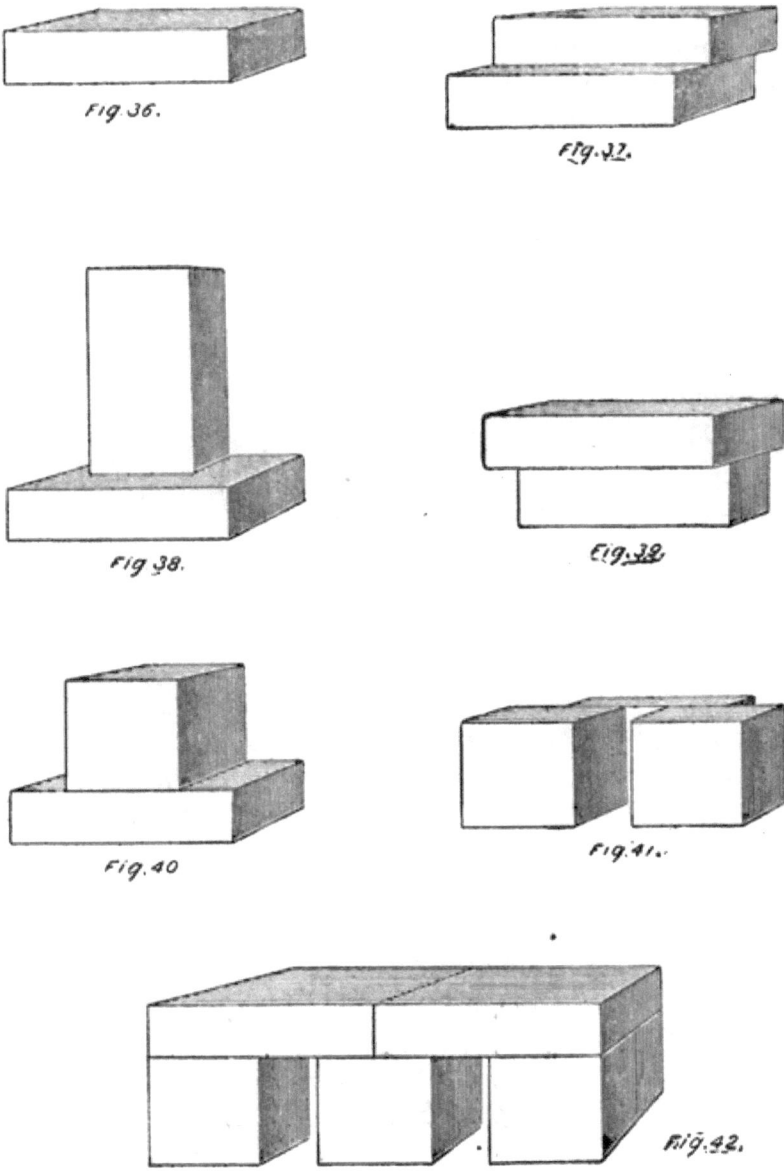

Fig. 36.
Fig. 37.
Fig. 38.
Fig. 39.
Fig. 40.
Fig. 41.
Fig. 42.

One cube, three oblongs, and three half-cubes.
Two cubes, one oblong, and three half-cubes.
Two cubes, two oblongs, and three half-cubes.
Two cubes, three oblongs, and three half-cubes.
Three cubes, one oblong, and three half-cubes.
Three cubes, two oblongs, and three half-cubes.
Three cubes, three oblongs, and three half-cubes.

All forms constructed in Block Building should be named by the pupils. Some true or fancied resemblance to a real object will suggest a name for nearly every form that can be made. The usual exercise may be varied pleasantly, by the teacher or some pupil in the class giving the name of an object, which all in the class try to construct, each according to his own idea.

A very few of the forms thus made are represented:—

PAGE 65.

Fig. 29, Two roofs.
Fig. 30, Little tent.
Fig. 31, Small house.
Fig. 32, Bridge.
Fig. 33, Barn.
Fig. 34, Light-house.
Fig. 35, Mill.

PAGE 67.

Fig. 37, Step.
Fig. 38, Clock.
Fig. 39, Bench.
Fig. 40, Bee-hive.
Fig. 41, Fire-place.
Fig. 42, Long bridge.

PAGE 69.

Fig. 43, Pigeon-house.
Fig. 44, Window.
Fig. 45, Locomotive.
Fig. 46, Engine-house.

## THE OBLONG BLOCK.

69

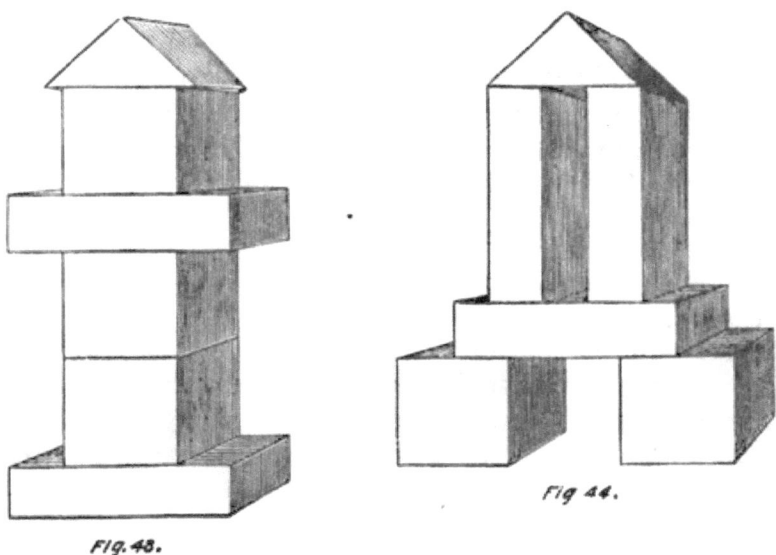

Fig. 43.

Fig. 44.

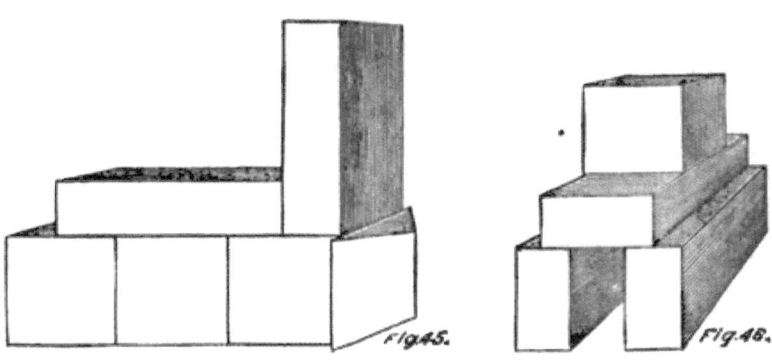

Fig. 45.

Fig. 46.

## STRAW STRINGING.

THE MATERIALS. Lemonade straws, so-called, may be purchased at drug stores, crockery stores, or other places, where they are kept for sale. A package containing a great number of straws about ten inches in length, will cost only a few pennies. They may also be gathered in quantities in a field of ripened wheat, or after the wheat has been harvested.

In order to prepare them for use, they should first be soaked in water for twenty-four hours; then removed and cut into appropriate lengths with a pair of strong, sharp scissors. Convenient lengths for stringing are as follows, viz.: One and one-half inch, one and one-fourth inch, one inch, and one-half inch. To measure the straws in cutting, place a rule on a thick book, extending it to the left, the required length, and by the side of this hold the straw with the left hand, cutting with the right hand.

---

### FIRST LESSON.

THE STRAW. In the new occupation now presented, both eye and hand are trained to accuracy. It is the child's first experience in handling the needle and thread. The material upon which he is to work is very fragile, and nicety of touch must be cultivated in order that the delicate straws may not be injured or broken. He must also learn to thread the needle, and to tie a good knot in the end of the thread.

The first exercise consists of an object lesson upon the straw. The teacher holds up before the class an uncut straw, and asks, "What is the name of this object?"

The pupils answer, "It is a straw."

*Teacher.* What is the use of a straw?

*Pupil.* It is to hold up the head of wheat or oats when it is growing.

*T.* Where does the straw come from?

*P.* It is a part of a plant.

*T.* Tell me something about a straw.

*P.* The straw is round.
   It is tall (or long).
   It is slender.
   It is hollow.

*T.* If I should try to bend the straw, what would it do?

*P.* It would break.

*T.* When anything can be broken easily, we say that it is *brittle.* Then what may we say of the straw?

*P.* The straw is brittle.

The straw is then placed in the hand of some pupil.

*T.* Is the straw light or heavy?

*P.* The straw is light.

*T.* Is it rough or smooth?

*P.* It is smooth.

Four pieces of straw, of the different lengths prepared for stringing, are now displayed.

*T.* What have I now? (Fig. 46, page 77).

*P.* Pieces of straw.

*T.* Count the pieces I hold in my hand.

*P.* One, two, three, four.

*T.* How do these pieces of straw differ from one another?

*P.* They are of different lengths.

*T.* Name some objects that resemble this piece of straw.

*P.* A blow-gun is something like it.

A piece of gas-pipe is round and hollow like that.

A pencil case is like it.

Some beads are like it.

*T.* What can be done, very easily, to anything that is brittle?

*P.* It can be broken easily.

*T.* Then how must we handle things of that kind?

*P.* We must handle them very carefully.

*T.* How must we try to use the pieces of straw when we take them in our hands for work?

*P.* We must be careful not to break or injure them.

The teacher may assist the pupil in forming proper answers to questions, by encouraging him to explain his ideas freely, and by aiding him in clothing his answers in correct language.

### SECOND LESSON.

STRINGING STRAWS OF THE SAME LENGTH. A brief review of the object lesson on the straw should preface the beginning lesson in Straw Stringing.

For stringing the straws a horse-hair or a needle and thread may be used, as the teacher may desire. Fastening the thread at the end, to prevent the straws from slipping off, is accomplished by tying the thread around the middle of a piece of straw. An ordinary knot in the end of the thread will pass through the hole in the straw. If needle and thread are to be employed, preliminary instruction and practice in threading the needle will be necessary.

When the pupils are in readiness for the lesson, convenient boxes containing a quantity of one-and-a-half inch pieces of straw are distributed to the class.

The teacher, holding a straw in a horizontal position in the left hand and a threaded needle in the right, in position for passing the latter through the former, dictates as follows to the pupils:—

Take one straw from your box. Place your straw and needle in the position in which I hold mine.

Pass the needle through the straw.

Remove the straw from the thread, passing it over the needle.

Now, as I count, place straws one at a time upon the thread.

(*Counting slowly*) One, two, three . . . . . ten.

Carefully remove all the straws from the thread, as I count.

(*Teacher counts as before.*)

String straws again, as I count.

(*Teacher counts faster than before*) One, two . . . . ten.

Remove the straws again from the thread.

(*Teacher counts as before.*)

String straws again, as I count.

(*Teacher counts still faster*) One, two . . . . . ten.

Remove the straws from the thread.

(*Teacher counts as before.*)

String as many straws as you can in a half-minute. (*Times them by watch.*)

While I count twenty.

From the time I say "begin" until I say "time."

Remove the straws from the thread.

Place ten straws upon the thread, one at a time, all counting.

(*Children count together*)   One, two . . . . . ten.

Remove the straws.

Place them carefully in the boxes.

The boxes are then closed and collected, and the thread and needles properly cared for by pupil appointed.

A second, third, and fourth lesson in Straw Stringing follow, upon the plan of the first, taking in consecutive order the straws of different lengths, one kind only at each lesson.

### THIRD LESSON.

COMBINING STRAWS OF DIFFERENT LENGTHS. For this lesson, boxes are distributed to the class, containing a convenient quantity of straws of the four different lengths prepared. Each pupil having been provided with needle and thread, the lesson proceeds upon the usual plan of dictation as follows:

Select from your box one straw of the longest kind.

String ten straws of that kind, counting.

Remove the straws from your string, placing those of the same length together.

Now select a straw of the next longest kind.

(Repeat former instructions until all the lengths of straws have been used.)

Select straws of the longest and next longest kind, ten of each.

String one of the longest; one of the next longest; one of the longest; of the next longest; of the longest, etc.

Remove the straws from the thread, counting.

This exercise is continued until each of the different lengths has been combined separately with each one of the other lengths of straws.

Since the same general method of presenting the various occupations to children of the first grade is employed in all alike, the specific instructions given for Block Building will be found equally applicable for other employments. Detailed description, which involves repetition, will therefore be omitted hereafter.

### FOURTH LESSON.

COMBINATIONS OF ANY NUMBER OF STRAWS OF DIFFERENT LENGTHS. So many variations are possible under the conditions of this head that occupation is furnished for many lessons.

The outline presented for this work is not intended to be exhaustive, but consists merely of brief hints. The details to which they point will afford to each instructor the best satisfaction if supplied by himself.

The first business of every lesson should be the proper preparation and distribution of all necessary materials.

For the lessons now presented, they are the same as those employed in previous work. For its continuation the following suggestions are given, viz.:—

*First.* Combining, in turn, one straw of each kind with two straws of each of the other kinds, alternating. Stringing by dictation, thus:—one long, two short, one long, two short, etc.

*Second.* Combining, successively, one straw of each kind, with three of each of the other kinds, alternating.

*Third.* Combining, successively, one straw of each kind with four of each of the other kinds, alternating.

*Fourth.* In the same manner combining two of each kind with two of all the other kinds, in turn. Stringing by dictation in each case, and removing from the thread by count.

*Fifth.* In the same manner combining,—

> Two and three.
> Two and four.
> Three and three.
> Three and four.
> Four and four.

*Sixth.* Making a string using the four lengths of straws, rotating thus: Longest, next longest, next longest, shortest, longest, etc.

*Seventh.* Making, in the same manner, a string of the four different lengths, beginning with the shortest.

*Eighth.* Stringing any definite number of each length of straws, combined with any definite number of all of the other lengths, rotating in any order desired.

*Ninth.* Making a miscellaneous string by dictation, thus: Shortest, longest, shortest, next longest, longest, longest, etc.

*Tenth.* Making an equilateral triangle by stringing three straws of equal length, then tying and cutting the thread. In the same manner making various kinds of triangles, also the square, pentagon, diamond, etc. (Figs. 47, 48 and 49, page 77.)

*Eleventh.* Inventive stringing.

## THE STRAW.

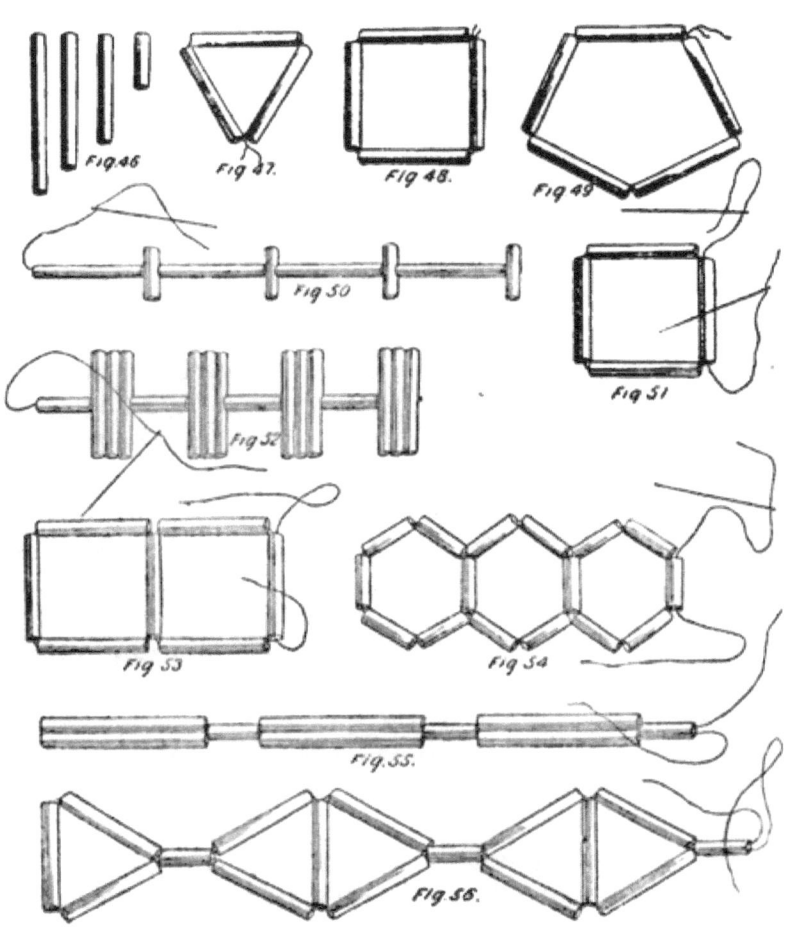

### FIFTH LESSON.

MAKING FORMS OF BEAUTY. Thus far all work in Straw Stringing has been done with the use of one needle and thread. The object of the present lesson is to indicate briefly, advanced steps in the occupation, requiring the use of two or more needles and threads. (See Figs. 53 to 63, pages 77 and 79.)

The usual box of straws, a long thread, and two needles having been supplied to each pupil, the teacher dictates as follows:—

Thread a needle upon each end of your thread.

Select from your box a straw of the shortest length; pass one of the needles through it, and slip it to the middle of your thread.

Next, string a short straw upon each of the needles and slip both to the middle of the thread.

Select another short straw. Pass both needles through it, one toward the right and the other toward the left. Slip this straw to the middle. What have you made? (Fig. 51, page 77.)

*Pupil.* A little window, or a picture-frame.

*Teacher.* Leave these four straws upon the thread. Again string a short straw upon each of the needles. Pass both needles as before through another short straw. Slip to the middle. What have you now made? (Fig. 53.)

*P.* A window with two panes.

*T.* Again select three short straws. Repeat the last operation. What have you made?

*P.* A short ladder.

This exercise may be continued indefinitely. It may be repeated also, using each of the other lengths of straws, or making combinations of different lengths.

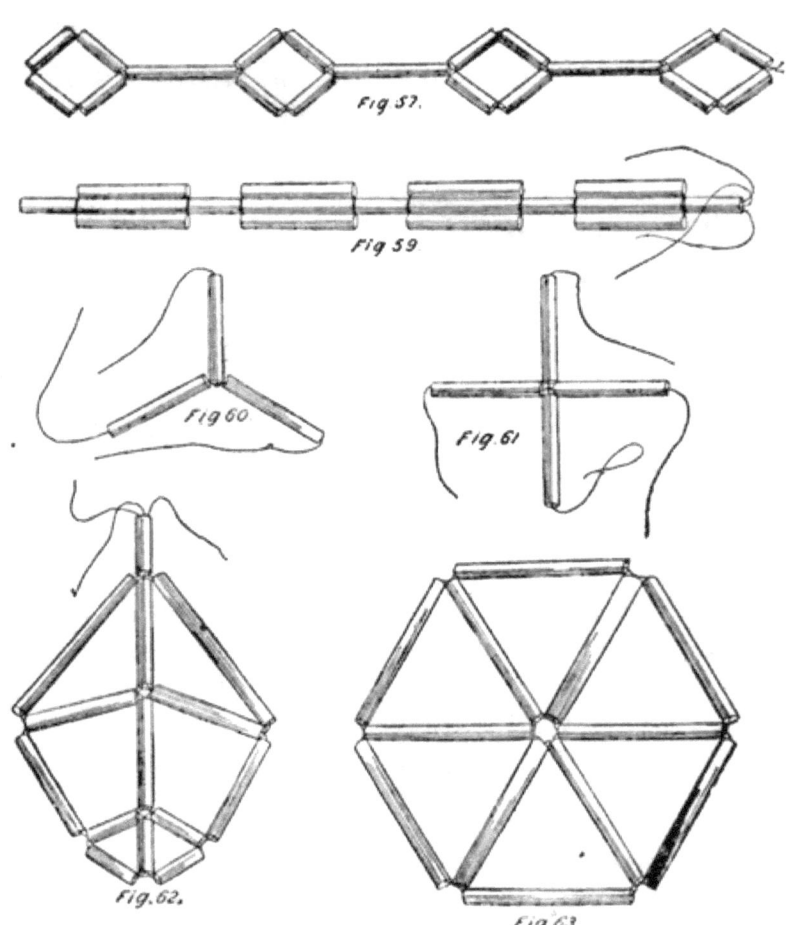

Except in the first practice-lessons, the straws need not be removed each time from the thread. Good specimens of the work each day may be preserved.

An indefinite number of lessons in Straw Stringing may be conducted upon the general plan of the exercise above described, varied as the details of each new form may require.

When sufficient time has been occupied in the use of two threads and needles, three may be employed and afterward four.

The work of Straw Stringing offers a large field for the exercise of the child's inventive faculty. Much ingenuity and skill will be displayed in the pretty forms produced each day in the time allotted for invention.

## BEAD STRINGING.

THE MATERIALS. The beads for stringing may be purchased of the dealers by the ounce, a few dimes getting enough for a large class. They should be of assorted sizes, colors, and lengths.

By taking care of them, seeing that they are not lost, and stringing and unstringing them over and over again they may be made to last for a long time.

For common use in stringing beads a needle and thread is the most satisfactory, but horse-hair or fine wire may be substituted for the sake of variety. These are easily procured at small cost.

PLAN FOR A SERIES OF LESSONS IN BEAD STRINGING. The foregoing plan for lessons in Straw Stringing will be found equally applicable to the occupa-

## BEAD STRINGING.

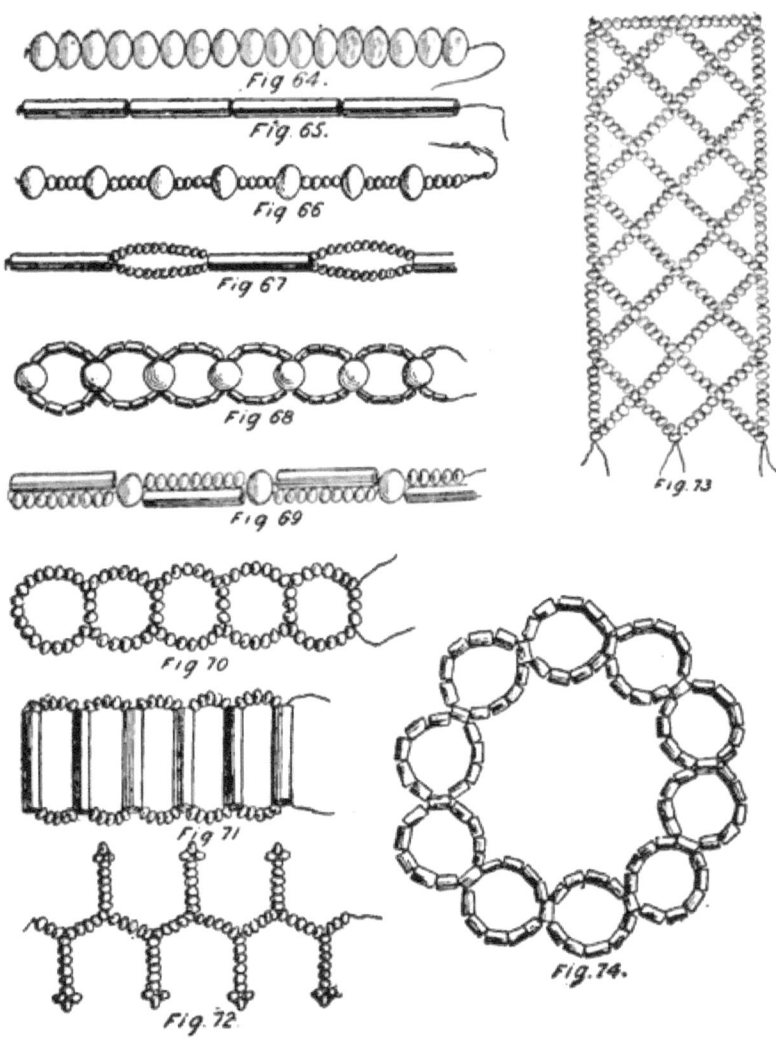

tion of stringing beads. A wider field of work is open, from the nature of which and the kind of material to be used, an exhaustive formula for lessons would be impracticable.

The variety of forms produced by combining beads of different shapes and sizes in different ways is almost infinite. (Figs. 64 to 74, page 81.) The occupation is interesting, and follows naturally in the line of advancement upon the pupils' former experience in the manipulation of straws.

The simple plan below suggested may be enlarged upon and varied by each instructor to suit his individual need.

1. Object Lesson,—subject, A Bead.
2. Stringing, with a single needle and thread, beads of the size and shape most nearly resembling the straws which they have been using. Stringing by dictation and by counting. Placing beads of the same kind together, as they are removed from the thread.
3. Stringing beads of any single shape designated, using one needle and thread.
4. Systematic combinations of beads of various sizes and shapes, using one needle and thread.
5. Making various forms, employing the long beads which resemble the straws previously used. Learning names of these forms, as Triangle, Square, Oblong, Pentagon, etc.
6. Using two or more threads with modifications and combinations of the forms produced by the use of one thread.
7. Invention in Bead Stringing.

## *LEARNING COLORS.*

THE MATERIALS. From the inexhaustible store of beautiful colors revealed to the eye in objects of external nature, the teacher will receive the best aid and material that can be furnished for teaching distinction of color.

For the beginning class exercise, a nucleus of material of the simplest and least expensive description will be needed, and around it there will speedily gather a supply that will more than equal the greatest demand.

If a color-chart be accessible certain advantages may be derived from its use.

The correct manipulation of a few good colors upon a palette will also constitute a valuable resource.

But, a few pieces of painted cardboard, or bits of ribbon of different colors, or pieces of cloth of any kind of material, or colored paper, or gelatine film, some painted blocks of wood, a handful of colored crayons,—*any* of these will answer the purpose sufficiently well,—*all* may be employed with advantage.

If the material be pieces of cloth or ribbon, paper, cardboard, or gelatine film, the pieces should be carefully cut of regular shape and equal size as far as possible. If blocks of wood, they should be smooth, truly cut, and neatly painted. If balls of worsted or yarn, they should be smoothly wound or covered.

However miscellaneous in its character the stock of material may be, it should be prepared invariably in a scrupulously neat, orderly, and symmetrical way, and should be replaced methodically, each kind in its appropriate box, at the close of every lesson.

A child who may, by chance, use any of the materials in a careless or untidy manner, will learn, perhaps, to value it, by deprivation, for a time, of the privilege of its use.

### FIRST LESSON.

BLUE. From a box containing pieces of ribbon of assorted colors, the teacher selects a blue piece, and holding it before the class asks the question, "What color is this piece of ribbon?"

The correct answer, "It is blue," will be given undoubtedly by one or more members of the class. To such as cannot name the color the teacher will give its name, and allow members of the class to name it in turn. Continual association of the color with its name will result speedily, in ordinary cases, in correct discrimination. The lesson continues thus:—

*Teacher.* What is the name of this color?
*Pupil.* It is blue.
*T.* Look about you and tell me what you can see that has this color.
*P.* I can see the sky. It is blue.
Mary's eyes are blue.
My dress is blue.
Some parts of the map are blue.
*T.* Name some things which you cannot see, which are blue.
*P.* Some flowers are blue.
I have a book that has a blue cover.
Uncle John's wagon is blue.
One part of the rainbow is blue.
Emily's sash is blue.

Each pupil is now permitted to select from the box a piece of ribbon of the same color as the one just discussed. In choosing the colors there will be found to be many different kinds of blue. The children are taught to designate these, as shades or tints of blue. A piece that is not strictly of a blue color, but resembles blue, may be called *bluish*.

The selections being made, and mistakes, if any, corrected by comparison with the pieces chosen by others, the teacher holding up a piece of ribbon sets the example to the class by saying, "My piece of ribbon is blue, like the sky.

The pupils, following in turn, say:—

"Mine is blue, like a forget-me-not."
"Mine is blue, like a violet."
"Mine is blue, like a robin's egg."
"Mine is bluish, like steel."
"Mine is blue, like a plum."
"Mine is blue, like the ocean."
"Mine is blue, like some parts of a peacock's feather."
"Mine is blue, like the stone in Kate's ring."

After this exercise, all of the pieces should be neatly replaced in boxes.

The class is excused with the request that each child bring to the class, at the next lesson, something that is blue in color. This will result in the collection of a varied assortment of bits of cloth, paper, glass, and leather, and all kinds of trinkets and beads,—a very museum of childish treasures.

When the lesson is reviewed, if the season be appropriate, each pupil may bring a blue flower to the class. Also, from a bouquet containing flowers of many different colors, the children may select all of the blue ones.

At another time, the boxes of beads, before used, may be distributed and the pupils instructed to select therefrom, all of the blue beads. For review, they may string them in various ways.

---

### SECOND LESSON.

PRIMARY AND SECONDARY COLORS. From the general plan of the lesson given upon the color Blue, lessons upon any other colors may be modeled. If preferred, such lessons may be continued uninterruptedly, until all of the chief colors have been exhausted as topics.

But if, after the three primary colors, Red, Blue, and Yellow have been used, a lesson be introduced for the purpose of showing how, by their combination, other colors may be produced, the result will not fail to greatly overbalance the slight effort required to produce it.

When the pupils have become so familiar with the three colors that they can name and select them unhesitatingly, they may learn to call them the "Primary Colors."

"What color is this?" asks the teacher, showing a blue piece of cloth, paper, or any material.

The pupils answer, "It is blue."

*Teacher.* What color is this (showing a red piece)?

*Pupil.* It is red.

*T.* And this (showing a yellow piece)?

*P.* It is yellow.

*T.* What are the colors, Red, Blue, and Yellow called?

*P.* They are called the Primary Colors.

*T.* Name the Primary Colors.

These colors being laid upon a palette, or any convenient smooth surface, the teacher can show, by the simple experiment of mixing any two of the paints together, how other colors can be formed by their combination. This is so easily done, and is so pleasant and interesting, that it cannot afford to be spared as an aid in teaching color.

The use of colored crayons upon the blackboard will be of much assistance in illustrating the principle of the lesson. (Figs. 75 and 76, page 89.)

After the production of the secondary colors, Purple, Orange, and Green, the name "Secondary Colors" may be given to the class. Each of these colors should be treated, in turn, in the same manner as the lesson on Blue.

### THIRD LESSON.

THE RAINBOW COLORS. The display of beautiful colors presented by the rainbow furnishes, of all the objects in nature, the most perfect illustration and subject for a lesson in color.

Having prefaced the lesson by a brief object-lesson upon the Rainbow as a whole, the teacher selects from the materials at hand, the seven Rainbow colors. These are displayed successively, to the class, who designate each color by its name as it is shown, thus: Violet, Indigo, Blue, Green, Yellow, Orange, Red. Also in the opposite order, Red, Orange, etc.

Then the colors may be placed upon the board with colored crayons, and named by the class in direct and reversed order, also miscellaneously, the teacher pointing. The pupils then count the Rainbow Colors, tell how many there are, and each is asked to name his favorite color and to mention something else that has the same color.

Then the seven colors are presented, one to each of seven pupils. As the teacher calls the colors in order, the pupils step forward and place them in proper succession.

The pupils may also be sent to the board, each in turn, to select a crayon, and thus represent the Rainbow Colors in their true order.

The production of the seven Rainbow Colors from the three Primaries, should also be illustrated. (Fig. 77, page 89.)

### FOURTH LESSON.

REVIEW EXERCISES AND TALKS ABOUT COLORS:—

1. Placing all the articles that the children have brought to illustrate color together upon a desk or table, and, from the miscellaneous pile, allowing one pupil to select all of the blue articles, another one all of the green, etc., placing each color by itself.

2. Comparing colors in different kinds of fabric, precious stones, flowers, feathers, smoke, clouds, etc., and show how the same color in different objects presents a different appearance.

3. Cultivating the taste by comparing different colors, discussing harmony and noting effects of various combinations and arrangements.

4. Showing how the Maker of all the beautiful things we see, has created numberless objects of ex-

## LEARNING COLORS. 89

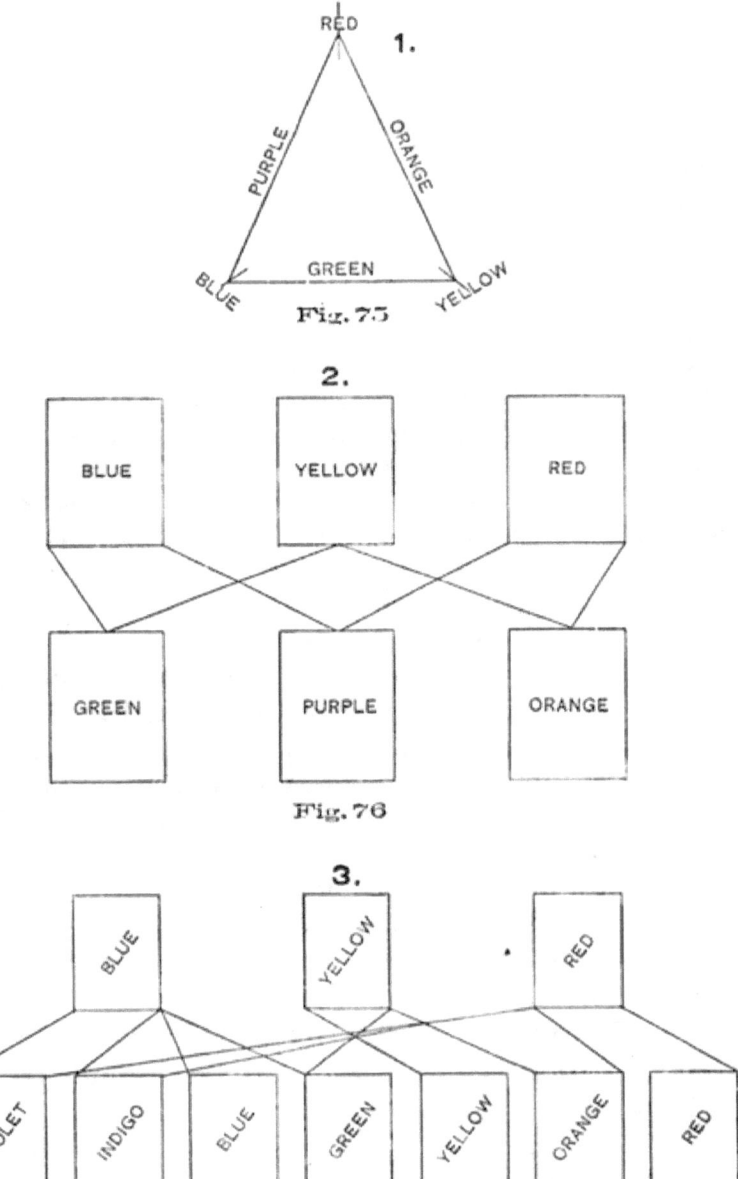

Fig. 75

Fig. 76

Fig. 77

quisite loveliness, whose colors, tints, and delicate shades we are not able to name.

5. Teaching that we ought to be thankful to Him for bright and beautiful colors, since the effect of bright colors is to make us feel cheerful and happy, and of dull colors to make us feel sad and gloomy.

## TABLET LAYING.

THE MATERIAL. The tablets may be made of common pasteboard. It is better, however, to make them of heavy strawboard, pressed and having a smooth surface. The best tablets are made of thick strawboard tarred. Several kinds of heavy board, suitable for the purpose, can be found at a bookbindery. Even a few old cloth book covers may be utilized for making them.

Although a great variety of tablets can be made, five kinds are quite sufficient for the general purposes of the work, viz., the inch square, the half square, made by cutting the inch square diagonally, the oblong one inch by two inches, the equilateral triangle one inch on each side, and the isosceles triangle, made by cutting the oblong through both diagonals, making one side two inches, and the other two sides about one and one-fourth inch.

The boards from which these tablets are made can be first painted in different colors, and then cut as usual, making them very useful for teaching colors.

PLAN FOR A SERIES OF LESSONS IN TABLET LAYING. A graduated series of lessons in Tablet Laying may be planned to follow in general detail the series presented for Block Building.

## TABLET LAYING.

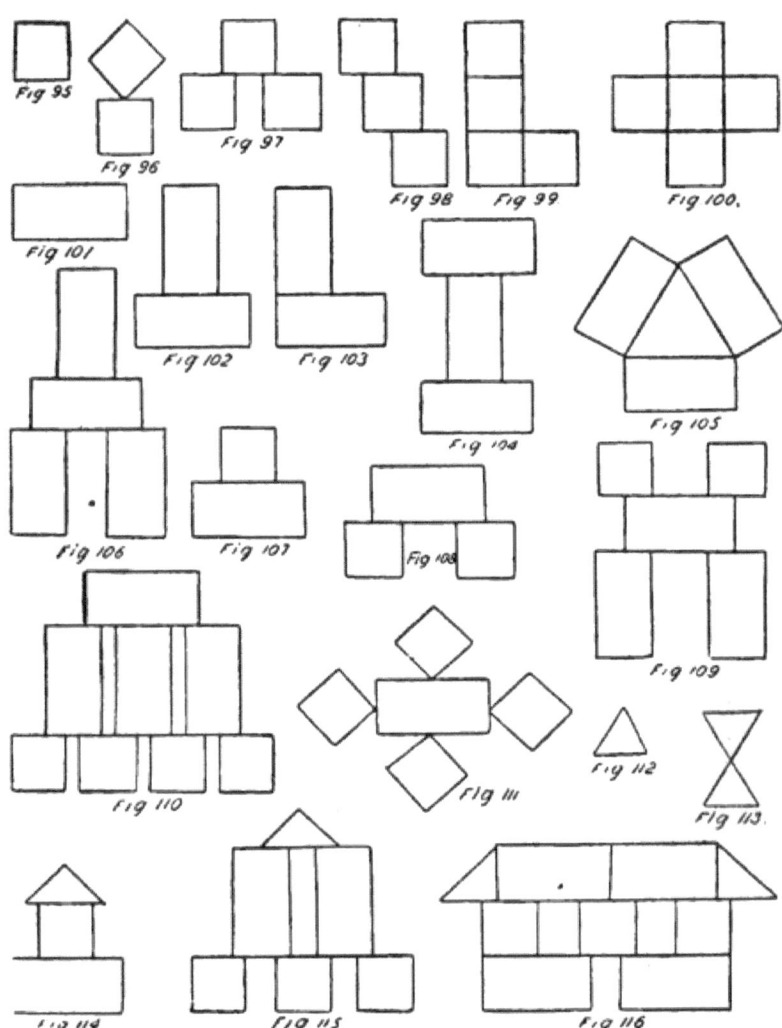

As in Block Building the introductory exercise consisted of an object lesson upon the cube, as the simplest representative of solid forms, so Tablet Laying is introduced by a similar object lesson upon the square, as the simplest representative of plane or surface forms.

It is designed to give, for this occupation, merely a brief outline, or series of guiding hints. The teacher will not fail to recognize the fact that in this, as in all of the topics heretofore treated, his best and greatest resource will lie in his own fund of originality and invention. The field is wide, fertile, and suggestive.

A simple plan is presented, subject to such revision as each instructor may deem necessary. (Figs. 78 to 116, pages 91 and 93.)

1. Object Lesson: subject, The Square Tablet.
2. Manipulating a single square tablet.
3. Using, in successive lessons, two, three, and four squares; making, by dictation, symmetrical forms and forms of life and beauty, naming each form.
4. Manipulating a single oblong tablet.
5. Using, in successive lessons, two, three, and four oblong tablets; making, by their combination, various forms, and naming them.
6. Combining the square and the oblong tablet, using one of each at the beginning, and advancing in succeeding lessons, with a definitely increased number of each, until at least four squares and four oblongs have been used in combination, working by dictation and naming the forms produced.
7. Manipulating a single equilateral triangle.
8. Combining the square, oblong, and triangle in the same manner as that described in Article 6.

## TABLET LAYING. 93

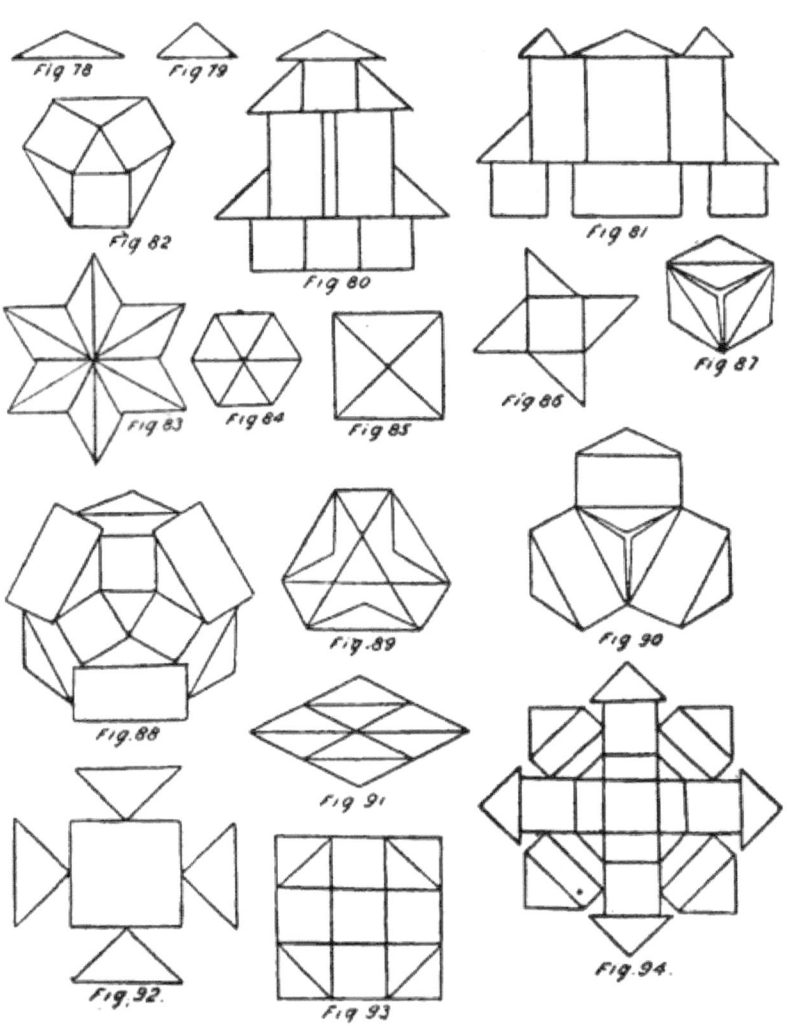

9. Manipulating a single isosceles triangle.

10. Combining all the forms in the manner indicated in Article 6.

11. Inventive work. Making and naming forms of beauty and symmetry.

LAYING COLORED TABLETS. After the eye of the child has become somewhat practiced in the observation of accurate and symmetrical forms made by the use of the plain tablet, a number of lessons combining ideas of both form and color will be found very profitable.

The laying of the tablet, in itself a delightful occupation to the child, acquires for him a double attraction when he is permitted to handle and combine in beautiful forms tablets of a variety of colors as well as shapes.

The entire field covered by the general plan for lessons in laying the plain tablets, may be reviewed, using the colored ones, and no fear need be entertained of the subject becoming wearisome or distasteful to the pupil by reason of much usage.

The opportunity for variation is so large that the occupation has constantly a new and beautiful aspect, and so by one of the pleasantest avenues possible the child is led to a nice cultivation of taste in color-harmony.

Every dictation exercise should be supplemented by a little time for invention. The pupil may be allowed, at first, the use of a limited number of tablets of but one shape and two colors. As the lessons proceed he is given an increased number of tablets of a variety of shapes and colors.

The scope of this work is almost inexhaustible, and a very wonderland of beauty is opened up before the child in the kaleidoscopic forms which his hands are able to produce following the guidance of his own fancy.

## PAPER FOLDING.

THE MATERIALS. Sheets of light pliable paper of fine texture and a variety of colors are readily procured of any dealer in such articles. The book-seller will furnish them in large quantities at small cost.

These sheets should be cut into four-inch squares. This work must be done very carefully, in order that the squares may be accurately symmetrical.

With great painstaking it may be done by hand with scissors or knife, but the best way is to take advantage of the facilities found in every bookbindery and newspaper office, and so insure a more perfect workmanship than it is possible easily to attain otherwise.

### FIRST LESSON.

THE PAPER. The teacher introduces the usual preparatory object lesson by holding up to the view of the class one of the pieces of paper prepared for this occupation, and saying, "What do I hold in my hand?"

The pupils answer, "A piece of paper."

*Teacher.* Tell me something about this piece of paper.

*Pupil.* It is square.

*T.* How many sides has it?

*P.* It has two sides.

*T.* Count the sides as I point.

*P.* One, two.
*T.* How many edges has it?
*P.* It has four edges.
*T.* Count the edges.
Pupils count as before.
*T.* How many corners has it?
*P.* It has four corners.
*T.* Count the corners.
Pupils count as before.
Individuals are asked to feel of the paper, and the teacher asks, "Is it rough or smooth?"
*P.* It is very smooth.
*T.* Tell me something else about it.
*P.* It is thin.
 It is glossy.
 It is flat.
A pupil is asked to take the paper in his hand, and the teacher asks, "Is it heavy or light?"
*P.* It is not heavy.
 It is very light.
*T.* Is it large or small?
*P.* It is quite small.
*T.* How long is each edge, do you think?
The pupils make a variety of estimates whose accuracy may be tested by measurement.
*T.* What color is the paper?
*P.* It is blue (green, yellow, pink, etc.).
*T.* What else can we say of the paper?
*P.* It is clean.
*T.* If we wish to keep it clean, how must we handle it?
*P.* Carefully and with clean fingers.
*T.* What may happen if we do not handle it carefully?

*P.* It may become soiled or torn.

*T.* Name some other things that we can do to the paper.

*P.* We can cut it, roll it, or fold it.

*T.* Is all paper like this?

*P.* No, all paper is not of that color.
> Some kinds of paper are stiff, hard and thick, and some are soft and thin.
>
> Some paper has lines upon it.
>
> Some has writing and some has printing and pictures.

*T.* Name some objects that have the same shape as this piece of paper.

*P.* A handkerchief has that shape.
> A shawl is square.
>
> A table cloth is of that shape.

*T.* Tell me some of the uses of paper.

*P.* Paper is used for making books.
> It is used to write upon.
>
> It is used for wrapping bundles.
>
> Money is made of paper.
>
> Paper is used for making kites.
>
> It is used for making paper dolls.

If desirable, when the class is somewhat farther advanced, this subject may be reviewed and enlarged upon. Then the children may learn by investigation of what materials paper is composed, how and where made, and where obtained.

Sufficient capital for the first lessons will be gathered from the simple facts which the child can easily and immediately understand and express, and by questions which he can satisfy by his present limited observation.

## SECOND LESSON.

FOLDING SIMPLE FORMS. At the beginning of the initiatory lesson in Paper Folding, when the pupil takes for the first time the paper into his own hands his attention should be called to its delicacy of color and texture. Insistence upon the point of cleanly and careful work should be constant, for the field is exceptionally well adapted for its illustration. Restriction from the use of the material, and therefore exclusion from the occupation, will induce an effort toward reformation in the most habitually untidy or persistently slovenly pupil.

At each lesson, before receiving the material, each child is asked in succession by the teacher:—

"What color do you prefer your paper to be?"

"I would like a blue one, if you please," or "a pink one, if you please," the child is taught to reply.

When delicate tints occur, whose names the child is unable to designate confidently, the former color lessons may be brought into requisition.

When all are provided with paper, the teacher dictates as follows: Place your paper squarely before you, with an edge parallel to the edge of the desk.

Point to the right hand edge. To the left hand edge. To the upper edge. To the lower edge. To the upper right hand corner. To the upper left hand corner. To the lower right hand corner. To the lower left hand corner.

Carefully take the lower edge and fold it from you toward the top, placing the edges neatly and evenly together. Crease it. Turn the paper around and open it. Tell me what you have made?

*Pupil.* A book. (Fig. 117, page 101.)

*Teacher.* Is it a real book?

*P.* No, it is a doll's book, or, a play book.

*T.* How many leaves has this book?

*P.* It has two leaves.

*T.* How many pages has it?

*P.* It has four pages.

*T.* Place the book, open, before you. Fold the edge nearest you neatly upon the upper edge as before. Crease it carefully. Open it and tell me what you have made.

*P.* A window. (Fig. 118, page 101.)

*T.* Place the paper, open, upon the desk. Turn one of the corners toward you. Place this corner upon the opposite one; fit the corners and edges exactly; fold and crease. What have you made?

*P.* A little shawl. (Fig. 119.)

*T.* Open the paper. Place it upon the desk and turn one quarter around. Fold the lower corner upon the upper, as before. What have you?

*P.* Another shawl.

*T.* Open the paper. Look in the middle, where all the creases meet, and tell me what you find.

*P.* A star. (Fig. 120.)

*T.* Place the open paper again upon the desk. Turn one of the corners toward you. Fold the lower corner from you to the center of the star and crease neatly.

Turn the opposite corner toward you. Fold to the center as before.

Turn the right hand corner toward you. Fold to the center. What have you made?

*P.* An open envelope. (Fig. 121.)

*T.* Turn the remaining corner toward you. Fold *from* you to the center of the star as before. What have you?

*P.* A closed envelope. (Fig. 122, page 101.)

At the close of the lesson each child who has performed the work acceptably may be permitted to take it home and display it as a specimen of his handiwork.

---

### SECOND LESSON.

SIMPLE FORMS (*continued*). Each pupil being provided with material, as before, the lesson is continued, as follows:—

*Teacher.* Make a little book. A window. A shawl. Another shawl. Look for the star in the center. Make an open envelope. A closed envelope. Notice where the corners, or flaps of the envelope meet. Turn that side of the envelope downward upon the desk.

Turn one corner toward you.

Fold this corner to the center and crease.

Turn your form and fold the opposite corner to the center in the same manner.

Turn each of the remaining corners toward you, and fold, as before, to the center of the star.

Turn your paper over and tell me what you see.

*P.* Four little squares.

*T.* What name may we give to this form?

*P.* A handkerchief case. (Fig. 123.)

*T.* Take the form in your hand. Notice the little pockets made by the four little squares. Press the four corners carefully together with the left hand. Pick out the pockets with the right hand. Crease

PAPER FOLDING.

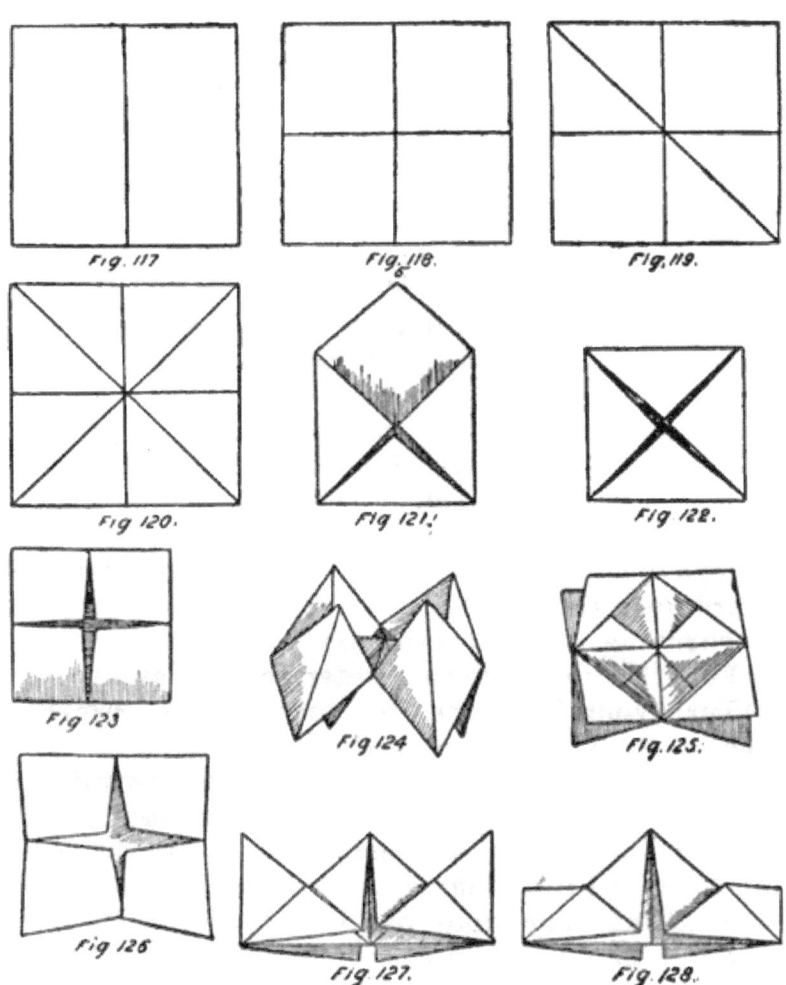

Fig. 117  Fig. 118.  Fig. 119.
Fig. 120.  Fig. 121.  Fig. 122.
Fig. 123  Fig. 124  Fig. 125.
Fig. 126  Fig. 127.  Fig. 128.

the diagonal fold in each neatly. Place your form upon the desk. What shall we call it?

*P.* A salt-cellar. (Fig. 124, page 101.)

*T.* Fold back the little flaps or pockets. What have you?

*P.* An open flower. (Fig. 125.)

*T.* Lay the open flower upon the desk. Press lightly with the finger upon the center. What have you?

*P.* A closed flower. (Fig. 126.)

A dextrous turn, easily understood, of the foldings already made, and a King's crown is the result. (Fig. 127.)

A slight change in this form produces the Queen's crown. (Fig. 128.)

Each of these pretty simulations of real objects of beauty is a delightful revelation to the child.

Any single form may constitute employment for an ordinary lesson, or more may be added if desired.

For each form the child begins with a new square of paper, folds in the succession described, the book, window, and other rudimentary forms mentioned, attaining by a final touch, in each case, a new and distinct form of beauty.

By the time this course of lessons in easy forms is completed, he will have attained considerable dexterity in handling the paper, and his simple *répertoire* in Paper Folding will consist in making, without assistance or dictation, any of the forms thus far described. He doubtless will make choice at different times, of different colors of paper, which will render his small display of forms quite attractive.

This rule should be carefully observed in every case, " Fold *from* you, and always *by opposites.*"

## WRITING.

This occupation in the First Grade, consists in learning to write from copies upon the chart or blackboard. Easy words which constitute the first reading lesson should be written by the pupil upon the slate and blackboard. As this branch of Manual Training has been long in the schools, no lessons are given.

## DRAWING.

This work consists of daily lessons in simple rudimentary drawing. Measuring off distances and drawing straight lines in different directions upon the slate are the usual employments for the first year. No lessons are given, as many books are published proposing specific instruction, but it is suggested that a great deal is to be discovered yet in properly teaching Drawing.

## GYMNASTICS.

Daily lessons in very simple free-hand exercises and marching should be given. Directions will be found in books published on this subject. It may be added, that schools well trained in gymnastics are noted for the order and happiness of the pupils.

## REVIEW LESSONS.

"PRACTICE MAKES PERFECT." The pupil who has received the benefit of instruction and practice in all of the employments thus far described, should have acquired considerable skill in the various manipulations. He will now, therefore, be quite prepared to make further advances in any of the kinds of work in which he has at any time engaged to a greater or less extent. This being the case, much opportunity should be given for reviewing former work. In this, however, the aim should be, not only to renew the child's knowledge of what he has learned before, but also to assist him in acquiring greater proficiency than he was able, at first, to attain. He will be able now to advance farther; will find new ground for pleasure and employment in fields which he had deemed exhausted; will perform the old work with new zest, having consciousness of the possession of greater power.

The review lessons should be conducted in every instance, upon a plan similar to, or adapted from, the original presentation of the subject. They should be conducted systematically. The subjects need not be reviewed in the order of their first presentation, but as each one is taken up for review, it should be used as a distinct lesson and continued in separate successive lessons until the interest seems to be in danger of flagging, or until the resources of the subject seem to the small workers to have become exhausted. After the child has been allowed the desired and, oftentimes, necessary change in occupation, opportunity may be given again for review of the same work.

In this, as in all branches of education, the review should not be conducted miscellaneously, hurriedly, or carelessly. In order to reap a harvest of better results than those obtained from the work when engaged in by the child for the first time, the lessons when reviewed should receive the same painstaking consideration as that which was originally accorded to them. The work should be prepared, the class received, conducted and dismissed, and the materials restored to their proper places, invariably in a perfectly orderly and systematic manner.

The teacher should put forth the greatest effort to make each and every lesson pleasant to each and every participant. His stock of resources and expedients for the attainment of this end should be inexhaustible. It may be laid down as an axiom, that good results follow the efforts of happy and contented workers.

The inventive power of the instructor should be continually active. Especially will he find this a necessity in review of familar subjects, in order that the pupil may find in the old employment newly presented, the same delight that an entirely new occupation affords.

To the faithful teacher, who appreciates fully the fact that skill is attained only through the medium of patient practice, and that practice in doing things often and well makes skillful doing a habit, the necessity of frequent and well-conducted reviews need not be further emphasized.

## Chapter II.

## SUGGESTIONS, LESSONS, AND METHODS OF INSTRUCTION IN MANUAL TRAINING.

THE PRIMARY SCHOOL—SECOND GRADE.

1. THE PUPILS. These pupils have usually attended school for one year. Their ages vary from six to seven or eight years.

2. THE LENGTH OF LESSONS AND AMOUNT OF WORK. For this grade the time occupied by each lesson should be of about the same duration as directed in plan for First Grade. The work may be presented upon successive or upon alternate days, as the requirements of the school and its program may demand.

The plans which follow embrace employment for many lessons. These lessons may be combined, divided, varied or supplemented, as each instructor may judge best. It is in Manual Training as in Language; the teacher must adapt his lessons according to good judgment.

3. THE STUDIES AND OCCUPATIONS. Reference to Appendix, Chapter II., will furnish an outlined plan of Studies and Occupations for this grade. They cover the same subjects as the first Grade, viz.: **Language, Numbers, Objects, Manual Arts.**

4. THE MANUAL ARTS. The occupations are as follows:—

Stick Laying.
Picture Cutting.
Scrap-Book Making.
Spool Work.
Paper Embroidery.

Braiding.
Writing.
Drawing (Inventive).
Gymnastics.
Reviews.

## *STICK LAYING.*

THE MATERIALS. The material required for Stick Laying consists of small, wooden sticks. Five cents will purchase a box of common wooden tooth-picks. This quantity will serve for a long time. They may be cut into various lengths.

LESSONS IN STICK LAYING. The preparation afforded the child by previous occupations of a similar nature does much to simplify the introduction and reception of this one.

In the lessons upon the cube and other solid forms, there was presented to his mind ideas of the three dimensions,—Length, Breadth, and Thickness.

In the square tablet and other plane forms, knowledge of the two dimensions,—Length and Breadth was gained. After these, by natural sequence, comes the linear form, or stick,—the representative of one dimension, Length.

A plan for a graduated series of lessons in Stick Laying:—

1. Introductory Object Lesson: subject, The Stick.
2. Manipulating *one* stick.

108  INDUSTRIAL EDUCATION.

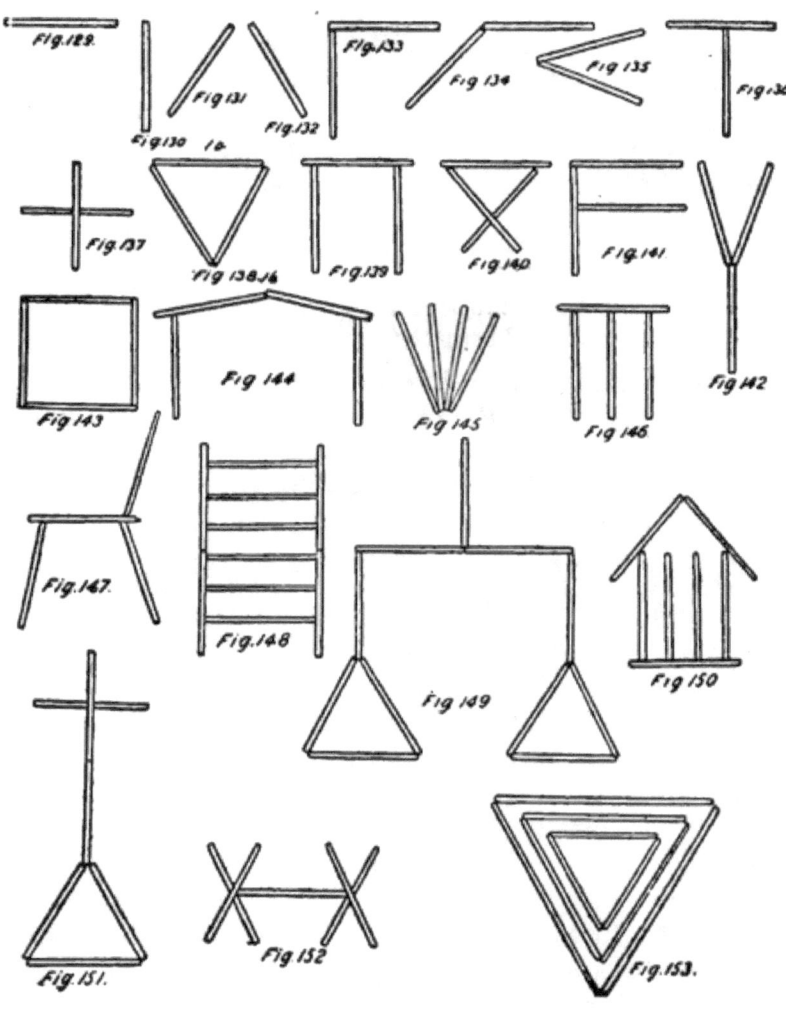

3. Making angles of various sizes, and other forms by combining *two* sticks, by dictation. Naming forms.

4. Making the triangle and other forms by combining *three* sticks, by dictation. Naming forms.

5. Making a variety of forms of symmetry, life, and beauty, by combining *four* sticks, by dictation.

6. Making forms by combination of any number of sticks, by dictation. Naming forms.

7. Using sticks of different lengths, and employing in separate succeeding lessons any desired combination, as regards length and number of sticks.

8. Reviewing the foregoing lessons, without dictation, giving much opportunity for invention. Naming forms.

Figures 129 to 159, pages 108 and 110, illustrate a few of the figures thus produced by Stick Laying:—

Figs. 129, 130, 131, 132, Rulers.
Fig. 133, Carpenter's square.
Fig. 134, Carpenter's rule.
Fig. 135, Compass.
Fig. 136, Milking stool.
Fig. 137, Cross.
Fig. 138, Triangle.
Fig. 139, Stand.
Fig. 140, Camp-stool.
Fig. 141, Flag.
Fig. 142, Hay Fork.
Fig. 143, Picture Frame.
Fig. 144, Gate-way.
Fig. 145, Fan.
Fig. 146, Grate.
Fig. 147, Chair.
Fig. 148, Ladder.
Fig. 149, Scales.
Fig. 150, Church-window.
Fig. 151, Cross.
Fig. 152, Saw-horse.
Fig. 153, Triangles.
Fig. 154, Rake.
Fig. 155, Whirligig.
Fig. 156, Easel.
Fig. 157, Kite.
Fig. 158, Coffee-pot.
Fig. 159, House, Barn and Dove-cot.
Figs. 160, 161, 162, Letters.

## *PICTURE CUTTING.*

THE MATERIALS. The multitude of illustrated publications which can easily be obtained at little or no cost, and the artistic pictures which the present favorite mode of advertisement furnishes, yield an abundant supply of material for this employment.

The single tool required by each child for the **work** is a pair of keen-edged, blunt-pointed **scissors**.

---

### LESSONS IN PICTURE CUTTING.

The occupation of picture cutting, always a favorite one with children, and usually prominent among their home amusements, gains rather than loses fascination for them when presented as a manual employment at school.

The study of pictures constitutes *per se* a most excellent field for the cultivation of the power of observation.

From the nature of the occupation a specified number of lessons cannot readily be planned. Each teacher must decide for himself how much time may be devoted to the work at its first presentation. It affords one of the pleasantest variations for an occasional review exercise.

In conducting the lessons it is suggested,—

1. That the child be taught first how to hold and manage the scissors correctly. One lesson-hour may be occupied in cutting long strips from a common newspaper, so that the little workman may acquire some deftness in handling his tool before he attempt intricate work.

2. That the child be allowed often, but not invariably, to choose the picture which he is to cut.

3. That the teacher strive to guide the child to a nice discrimination in his choice of a picture, and to cultivate his taste for, and understanding of, a high order of beauty in form and color.

4. That with this end in view the teacher and child talk much about the picture, and study its points together.

5. That the children assist in selecting pictures and furnishing material for the occupation.

6. That, in some instances, the picture be cut by following the outline of some figure or figures represented in it, and that in others the marginal outline of the picture, taken as a whole, be followed.

7. That the child, having made his choice of a picture, shall not be allowed to change, but shall finish cutting it before he be permitted to take another. That he be taught to follow the outline carefully and patiently in all of its details, and that the child, who, in impatience or carelessness, shall tear his picture, or cut it badly, shall be denied the privilege of the occupation until he shall manifest his desire to renew it by the promise of painstaking effort.

8. That the child be taught this principle of economy, viz., that *abundance or cheapness of material does not justify or excuse wastefulness in its use.*

9. That at the close of each lesson in Picture Cutting the child be required to put his work carefully away in a neat and orderly manner in its proper place, and to pick up and neatly dispose of all the litter which he has made in cutting and trimming his picture.

## SCRAP-BOOK MAKING.

**THE MATERIALS.** The scrap-book may be made of cloth or of paper. The pictures may be pasted upon separate leaves, and these leaves be fastened together to form a book. When finished the outer cover of this book may be decorated with some pretty picture, and the back neatly fastened with bows of ribbon.

Another plan is to make the book, and paste the pictures into it afterward. Still another is to make use of the common form of scrap-book kept by booksellers.

A smooth, clean flour-paste or common gum arabic may be used for pasting.

Clean, soft cloths or brushes should be provided, for the removal of every trace of paste or gum from the page, or from the edges of the picture.

---

**LESSONS IN SCRAP-BOOK MAKING.**

From cutting pictures, the step is natural and easy to mounting them.

If a number of separate leaves, or books, be provided, several of the pupils may engage in the work at the same time. If not, they can work, in turn, upon the same book. While awaiting his turn, each may cut the picture which he is to mount. The pictures should be pasted upon one side of the page only.

In order to derive the best benefit from this employment, the child should be instructed,—

*First*—To prepare his pictures for the scrap-book by cutting them carefully.

*Second*—To begin the work of each lesson with clean hands and well-ordered material.

*Third*—To arrange his pictures upon the page tastefully, effectively, and grouped as far as possible with reference to the subject which they represent or illustrate.

*Fourth*—To begin, when possible, at the upper left hand corner, and work across and down the page.

*Fifth*—To use as little paste or gum as possible, and to make very great effort to keep his hands, clothes, and all material, clean and neat.

*Sixth*—To work patiently, to handle the material delicately, and to strive to attain perfect workmanship.

*Seventh*—To clear away all litter when the lesson is done, and to have a proper pride in excelling in the important particular of performing his work in a clean and orderly manner.

## SPOOL WORK.

THE MATERIALS. The material for this occupation consists of a quantity of bright-colored worsteds, wools, or yarns. These, if bought in such quantities as would be required for class use, can be procured from local dealers at very moderate rates. Such materials are usually sold in bunches, each one ounce in weight. The worsteds should be of the kinds commonly known as split, single, and double zephyr, and the yarn such as is employed in common knitting. The greater the variety procured in color, the greater will be the attraction of the work for young children.

The tools required are, first, a sufficient number of empty spools of medium size to supply each member of the class with one. These, the home work-baskets will furnish. Into one end of each of the spools are driven four small, strong pins or tacks, equidistant from each other, and near to the edge of the hole in the spool. A few cents will purchase a large quantity of these small pins, from any hardware dealer. (Fig. 164, page 117.)

Second, a large pin or a common darning needle for each pupil. (Fig. 163.)

For practice in spool work the yarn may be used; for a nice piece of work to be completed, the worsted of either kind mentioned. More than four pins may be used, in each spool, if desired.

### LESSONS IN SPOOL WORK.

The first lesson in this occupation should be employed in teaching the pupils how to wind the worsted skillfully from the skeins into smooth, round balls, and, at the same time, to wind it very loosely. The latter object is usually attained by winding over the fingers of the hand holding the ball, and slipping it off at convenient intervals.

The child's first piece of spool work may be "set up," or started, by the instructor. After the first piece he must learn to begin the work for himself. He should be permitted to select worsteds of the color or colors which he wishes to use.

The work is commenced by taking a spool in the left hand, and, with the right, placing a little loop of yarn over each one of the little pins. The short end of the yarn is then passed through the hole in the spool and drawn out at the opposite end.

The work being thus prepared, the child is taught to hold the spool between the thumb and fingers of the left hand, and to guide the yarn easily with the same hand, carrying it by a natural movement around the pins toward the left, turning the spool as required. At the same time to pass the large pin through each loop from above, lifting it above the yarn and over the head of the corresponding pin. To hold the large pin in the right hand, and with the same hand to pull the work firmly down to its place upon the pins after each stitch, by means of the yarn projecting through the hole in the spool. (Fig. 165, page 117.)

In a short time the work itself will grow long enough to extend through the spool. When its length becomes inconvenient, it may be wound into a neat ball.

When a sufficient quantity of the cord has been made, the child may be taught to sew it upon a piece of firm cloth or paper, of circular shape, to form a mat; commencing this work in the center and fastening the cord, by long stitches upon the back, in spiral form, until the mat is as large as desired. (Fig. 166.)

To accomplish spool work well the pupil must endeavor,—

*First.* To handle his large pin so deftly and so carefully that he shall be in no danger of receiving a wound from its sharp point, or of wounding his neighbor.

*Second.* To hold the work handily, and not to unwind an unnecessary quantity of yarn from the ball.

*Third.* To watch his work closely and avoid dropping or omitting stitches, and to learn to take

## SPOOL WORK. 117

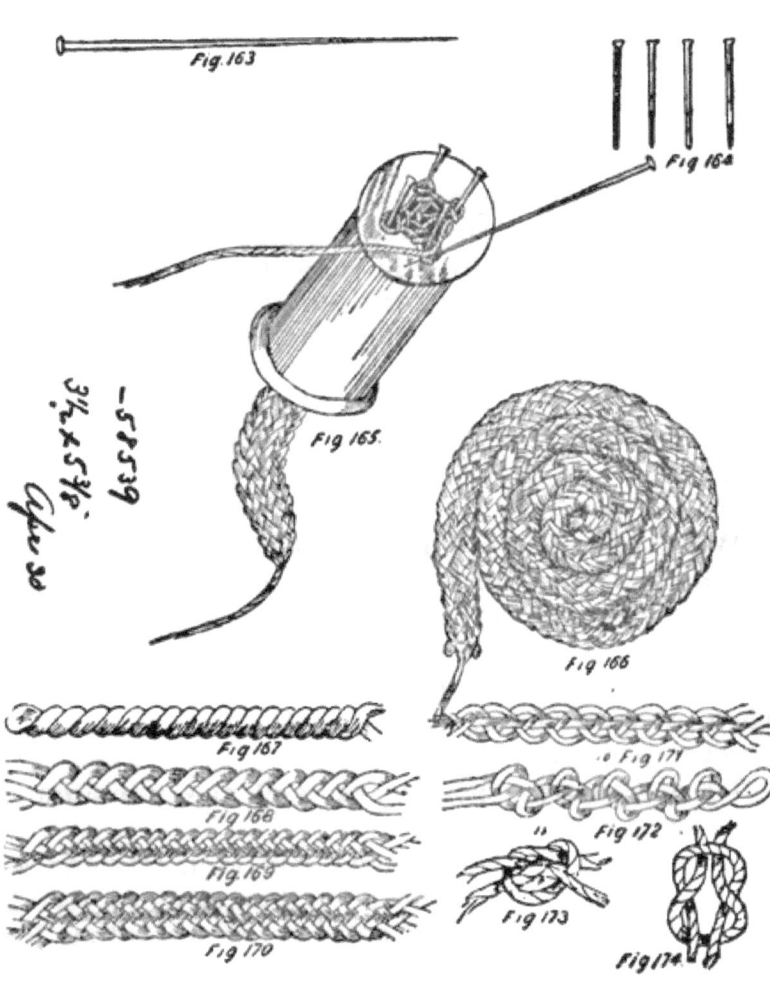

up dropped stitches and to take out work and remedy a mistake when one is made.

*Fourth.* To exercise good taste in the selection of colors, that the combination may be pleasing when the work is completed.

*Fifth.* To keep the work tidy and clean, and to put it away nicely after each lesson.

A small piece of paper or card, upon which his name is neatly written, should be furnished to each pupil. Before submitting his piece of work at the close of each lesson, this small label is pinned upon it, to insure recognition and avoid confusion when the work is next distributed.

---

## *PAPER EMBROIDERY.*

THE MATERIALS. The materials requisite for the occupation of embroidery on paper consist of a quantity of blotting paper, worsted of a variety of colors, and worsted needles.

The blotting paper is easily procured at all bookstores, at little or no expense. It is prepared for use in classes, by cutting neatly into three-and-a-half or four-inch squares. Upon each of these squares is drawn lightly, with lead pencil, a symmetrical design in straight lines. If the little workers are not able to do this for themselves, the designs may be drawn by the teacher, with, if desired, the assistance which older pupils are always glad to render.

The worsted and needles are procured as described under the head of Spool Work. The kind of worsted known as split zephyr usually finds greatest favor for Paper Embroidery.

PAPER EMBROIDERY. 119

## LESSONS IN PAPER EMBROIDERY.

An object lesson upon the needle, the thimble, the worsted, or yarn, and upon the blotting paper, will not be amiss in introducing this employment.

Practice in winding worsted and yarn into good balls should be continued.

Before entering upon the work of embroidery, the pupil must learn to thread the worsted needle, since he has now reached his first experience in its use. This is accomplished by holding the needle firmly in the right hand, with the point outward; folding a short end of the worsted over the needle and drawing it down firmly; then pulling the needle out of the flat loop thus made; next reversing the needle and pressing the eye upon the smooth, flat loop; finally, with the right hand drawing the worsted through the eye. (Figs. 251, 252, page 143.)

For the first embroidery lesson, the design should be simply the square (Fig. 175, page 119); for the next, the square with diameter (Fig. 176); for the next, the square with diagonals (Fig. 177); and for the next, the square with both diameters and diagonals (Fig. 178).

These are followed by more intricate designs (see illustrations, Figs. 179 to 191, pages 119, 121), and when the pupil has become quite familiar with the work, much time should be given for invention. When all of the pupils of a class are embroidering the same symmetrical design, the work may be done by dictation. Following symmetrical designs, forms of life and beauty may constitute the pattern. The scope for invention in this department is unlimited. Several colors may be used, if desired, upon the same design.

## PAPER EMBROIDERY. 121

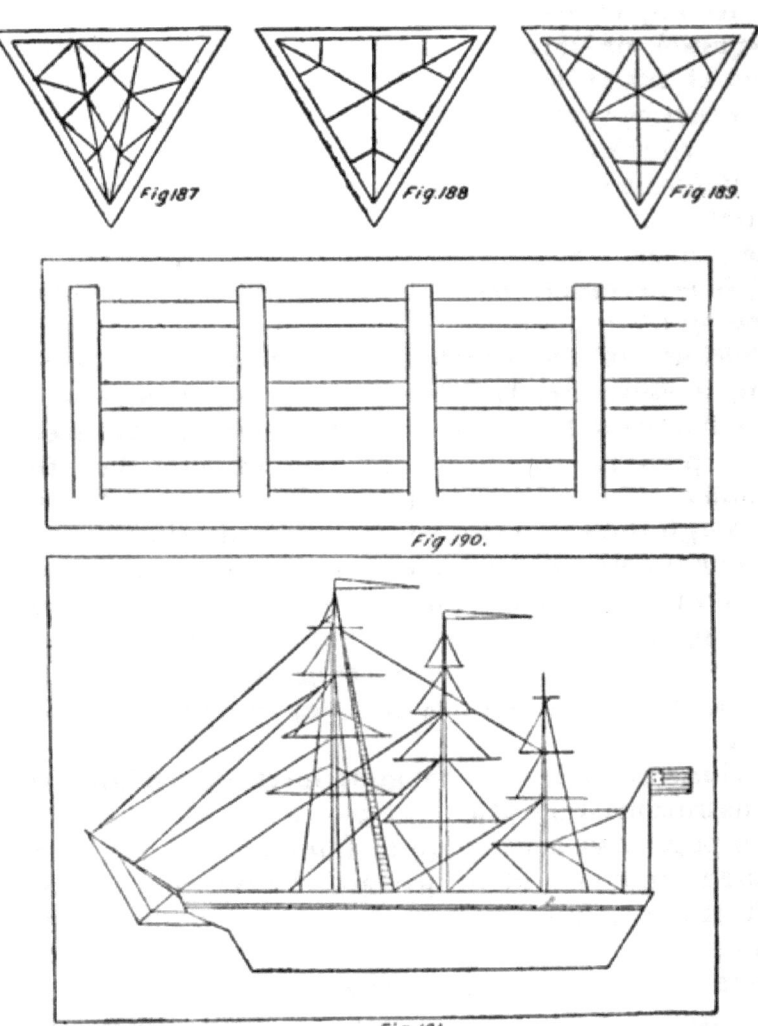

Fig. 187.    Fig. 188.    Fig. 189.

Fig. 190.

Fig. 191.

The pupils being supplied with proper materials and implements for work, are instructed, first, to make a puncture in the paper with the needle, in every place where the lines of the design cross or meet. Next to pass the threaded needle from the under to the upper side of the paper, through one of the punctures, and to cover the lines of the design with the worsted, by passing stitches from one puncture to another.

In this employment the child should be taught to observe:—

*First.* That he draw the worsted carefully through the puncture, so that he will not tear the blotter.

*Second.* That he strive to be very patient and careful in removing the tangled knots that will form sometimes in the thread.

*Third.* That he be as economical as possible in the use of the worsted, by making the short stitches on the lower or wrong side of the paper, and keeping the long ones upon the upper or right side.

*Fourth.* That he exercise good taste and judgment in selecting harmonious colors of worsted for his embroidery.

*Fifth.* That he endeavor, always, to keep the work clean and tidy.

*Sixth.* That he label his work properly, if unfinished, at the close of each lesson, and place all materials in order.

## *BRAIDING.*

THE MATERIALS. Any kind of common cord, or such yarns and worsted as are employed in spool work, fill all the necessary requirements for the occupation of Braiding.

## LESSONS IN BRAIDING.

The employment of Braiding, which covers sufficient ground for several pleasant and profitable lessons, consists in learning to make cords by twisting and braids by plaiting a given number of strands.

In learning to make a smooth, even cord or plait, accuracy of both eye and hand are cultivated.

An outline is given for a number of simple lessons: —

1. Object Lesson: subject, The Material (silk, worsted, etc.).

2. Learning to hold the strands firmly, and to twist or braid evenly, by counting or dictation.

3. Learning to twist a smooth, even cord of two three, four, or more strands. (Fig. 167, page 117.)

4. Learning to plait a braid of three stands. (Fig. 168.)

5. Learning to plait a braid of four strands. (Fig. 169.)

6. Learning to plait a braid of five, and finally of any number of strands. (Fig. 170.)

7. Learning to make a braid, plaiting the strands very tightly.

8. Learning to make a braid, plaiting the strands very loosely.

9. Twisting cords and plaiting braids, combining strands of different colors.

10. Learning to slip a braid upon one of its strands. (Fig. 171.)

11. Learning to untwist a cord and to unplait a braid neatly.

12. Learning to tie knots of various kinds, as, the weaver's knot, etc. (Figs. 173 and 174.)

13. Invention with various materials.

## WRITING.

The work of writing upon the slate and blackboard is continued as in the First Grade. Writing in the first grades of Tracing and Copy Books, using pen and ink, is also introduced.

## DRAWING.

Daily lessons in drawing consist in measuring off distances and drawing straight lines in different directions, as in the First Grade, also in making combinations of straight lines, drawing angles of various kinds, curved lines, and simple objects, and in inventive drawing. The slate and blackboard are used.

## GYMNASTICS.

Additional exercises in free gymnastics and marching, to those given in the First Grade, will be needed.

[NOTE.—The remarks made under Writing, Drawing, and Gymnastics in Part III., Chap. I., should be referred to.]

# Chapter III.

## *SUGGESTIONS, LESSONS, AND METHODS OF INSTRUCTION IN MANUAL TRAINING.*

### THE PRIMARY SCHOOL—THIRD GRADE.

1. THE PUPILS. As a rule these pupils have completed the work of the first two years of school. Their ages range, usually, from seven to eight or nine years.

2. THE LENGTH OF LESSONS AND AMOUNT OF WORK. The directions given under this head in Parts First and Second of this chapter, will be found to be applicable in the Third Grade as well.

3. THE STUDIES AND OCCUPATIONS. These are detailed in Appendix, Chapter II. The subjects embraced are Language, Numbers, Objects, Manual Training.

4. THE MANUAL ARTS. The following list embraces the occupations fitted for this grade:—

    Perforated Cardboard Embroidery.
    Slat Plaiting.
    Mat Weaving.
    Writing.
    Drawing.
    Gymnastics.

## PERFORATED CARDBOARD EMBROIDERY.

THE MATERIALS. A quantity of worsted of the kinds known as split zephyr or Shetland wool, a number of fine worsted needles, and some sheets of perforated cardboard, constitute the materials necessary for this occupation.

The worsted and needles are procured as explained in Chapter II., Part Second (Spool Work). The various kinds of cardboard may be bought in the same way of local dealers at the small expense of four or five cents per sheet. For practice work by beginners, the cardboard should be cut into four or five inch squares.

A sufficient number of good thimbles to supply a class can be purchased cheaply of the same dealer. A few pairs of good scissors and shears should be purchased at a hardware store.

### LESSONS IN PERFORATED CARDBOARD EMBROIDERY.

The cardboard, by reason of its equidistant perforations, presents a most excellent foundation for the embroidery of symmetrical figures. This work is performed in a manner similar to that employed in Paper Embroidery before described, the difference lying in the fact that an outline design is necessary in embroidering on paper, while the figure is made upon cardboard by counting a certain number of perforations in any required direction and connecting them with stitches of various lengths.

The first simple work of this occupation may be performed by dictation and counting, often when all of the members of a class are engaged in the same work, by concert counting. A plan for lessons in Cardboard work may be arranged as follows, subject to such variations as the teacher may wish to make. (Figs. 192, 193, page 129.)

1. Object lessons on materials used.

2. Review practice in winding worsteds, using thimble and scissors, and threading and handling needle.

3. Cutting the cardboard by counting perforations and following exact lines with the scissors.

4. Learning to make a simple stitch horizontally, connecting two adjacent perforations, working both from right to left and from left to right.

5. Making a horizontal stitch omitting one, two, three, and, finally, any given number of perforations.

6. Making vertical stitches in same manner, embroidering both from and toward the worker.

7. Making oblique stiches in the same manner.

8. Making a single cross stitch in four adjacent perforations.

9. Making two, three, finally, any given number of cross stitches horizontally adjacent, observing to cross all in the same way.

10. Making any given number of cross stitches vertically adjacent, crossing all in the same way.

11. Making any given number of cross stitches obliquely adjacent, making rows to the right and left oblique, crossing all in the same way.

12. Making successively any given number of cross stitches in horizontal, vertical, and oblique rows, the stitches not being adjacent, but one, two, finally, any number of perforations being omitted between the stitches in each instance.

13. Making larger symmetrical cross stitches in any four equidistant perforations not adjacent.

14. Making a simple border around a four-inch square of cardboard, using in combination any of the stitches above mentioned.

15. Inventive work in cardboard embroidery; employing any of the stitches learned.

## SLAT PLAITING.

THE MATERIAL. The slats used for plaiting are those made for tapers, and can be purchased of the dealers at drug stores, crockery stores, and other places.

They are made of some tough, flexible wood, and vary in length from four and a half to ten inches. The slats of these lengths may be cut very easily if other lengths are desired. They are about three-eighths of an inch in width and one-sixteenth of an inch in thickness.

The cost of this material is nearly nominal. As an occupation Slat Plaiting is one of the best in the Manual curriculum.

### LESSONS IN SLAT PLAITING.

The first work of the pupil after receiving the material consists in various manipulations, first of a single slat, then two, and then three slats, since at least four slats are necessary before Slat Plaiting proper may begin. As the lessons proceed the child will learn by practice and instruction how the whole structure of any figure made by interlacement of slats must depend upon the *bind* produced by the

## CARDBOARD EMBROIDERY. 129

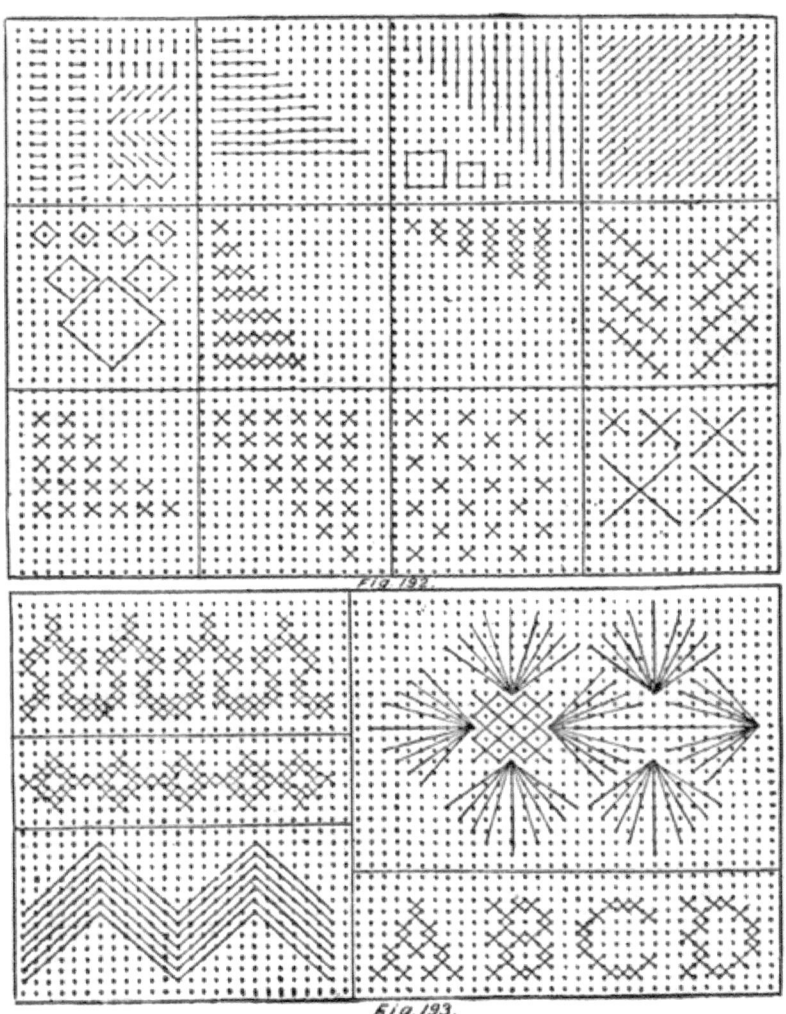

Fig. 192.

Fig. 193.

proper laying of the slats in relation to one another in the correct insertion of the final or binding slat. Until this slat is properly inserted, the removal of the form as a whole from the plane whereon it has been constructed, is obviously impossible.

This work requires much patient care and painstaking. Many failures must be encountered before the little hand will become skillful in plaiting the shifting slats into stable forms. Proficiency only can be attained by making a simple beginning of easy forms composed of a few slats, then by slow degrees making more complicated ones requiring the use of many slats. The general order of a series of such lessons is briefly shown in the following scheme:—

1. Introductory Object Lesson: subject, The Slat.
2. Manipulating one slat, as in Stick Laying.
3. Manipulating two slats, naming forms made.
4. Manipulating three slats, naming forms made.
5. Plaiting four slats, naming forms. (Figs. 194, 195, 196, page 131.)
6. Plaiting five slats, naming forms. (Figs. 197, 198, 199.)
7. Plaiting six or more slats, naming forms. (Figs. 200–205.)
8. Counting and naming angles of various forms constructed. (Page 131.)
9. Counting sides and corners.
10. Constructing figures and changing them to other forms.
11. Interlacing slats of various lengths.
12. Making forms of use and beauty. (Figs. 206–210, page 133.)
13. Invention in Slat Plaiting.

## SLAT PLAITING.

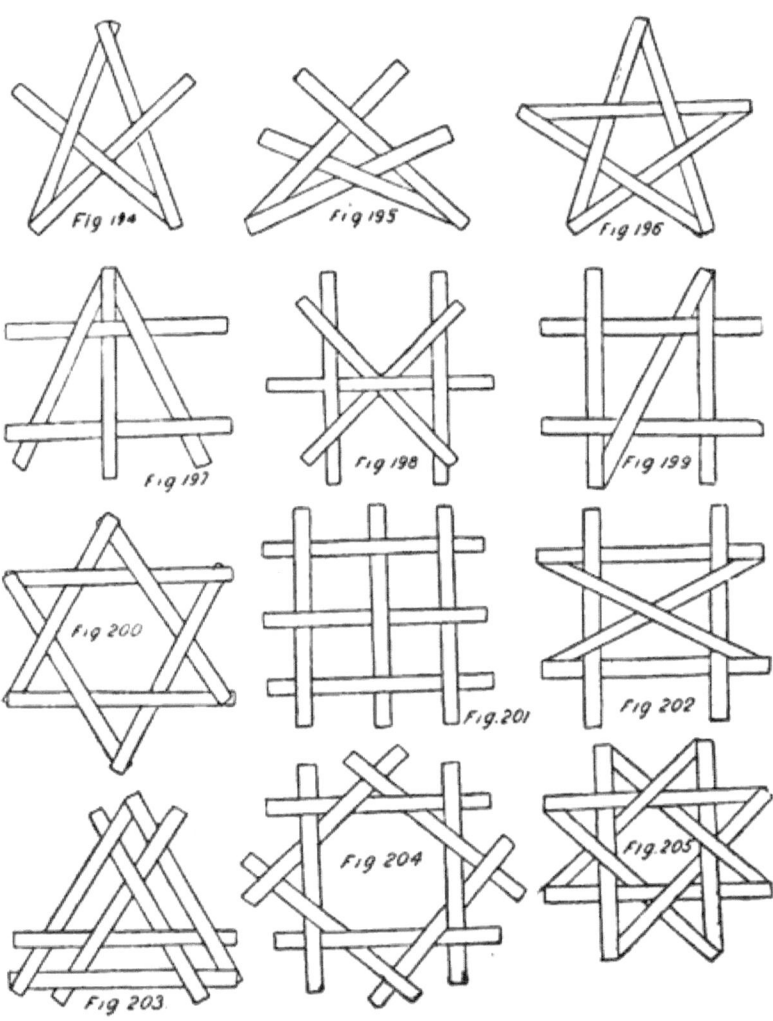

## MAT WEAVING.

THE MATERIALS. The materials for this occupation consist of mats and fringes of paper, and the steel weaving needle.

The mat or weaving sheet is a piece of fine glazed paper, in size about eight by six inches. The body of the mat is cut lengthwise into strips which, however, remain undetached from the margin or border, which is uncut. (Fig. 212, page 135.) The strips are cut of different widths in different mats; also a number of widths in the same mat. The fringe (which may be called the *filling*, as the mat the *warp*) is a sheet of the same dimensions as the mat, and is cut into strips or fringes lengthwise in the same manner, except that at one end the strips are quite severed from the body of the sheet and from each other, so that they may be readily detached, one by one, as needed for use. (Fig. 211.)

Both the mats and the fringes are of a great variety of colors.

The needle consists of a steel shank about eight inches in length, having a flat spring at one end. (Fig. 213.) The use of the spring is to hold the strip of fringe firmly during the process of intertwining it with the meshes of the mat.

All of these materials can be procured at small expense from dealers in Kindergarten materials.

---

**LESSONS IN MAT WEAVING.**

In all of the lessons in plain Mat Weaving the plan of conducting by dictation will be found most convenient and satisfactory.

SLAT PLAITING.

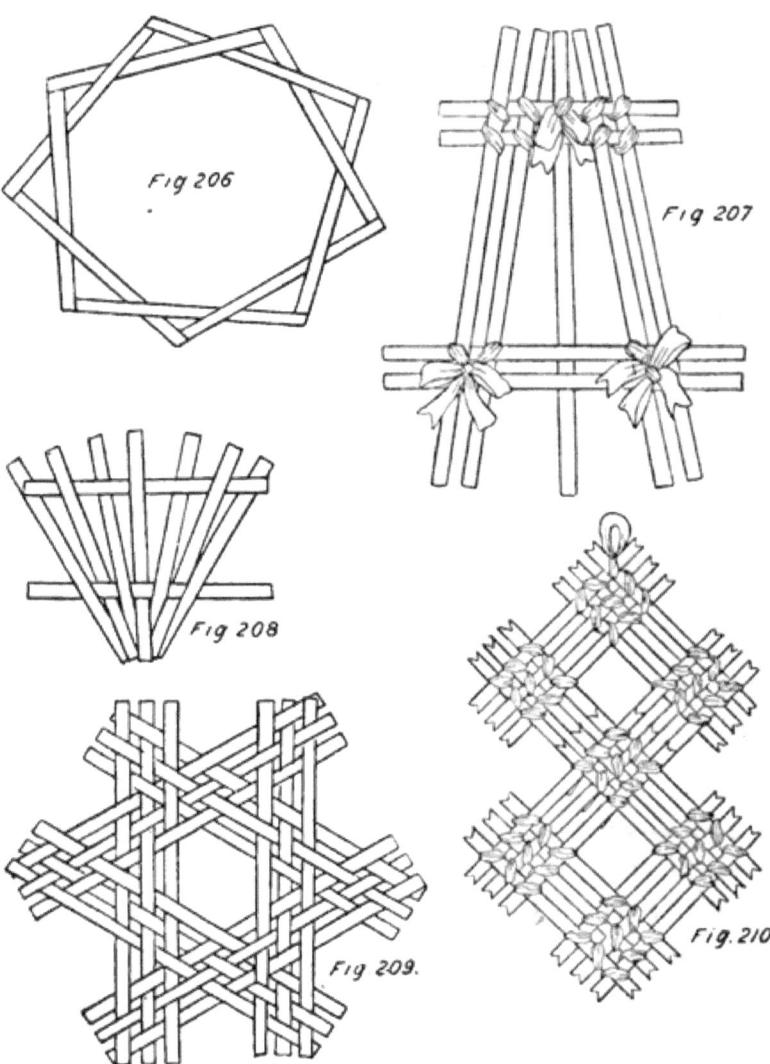

Fig 206.

Fig 207.

Fig 208.

Fig 209.

Fig. 210.

In every piece of weaving, two colors at the least may be combined; more than two colors may be employed when the pupil has made sufficient advancement. The work, therefore, will call into use the previous knowledge of the child in the special province of color, and will afford an excellent occasion for the exercise of his skill in combining colors effectively. His sense of the beautiful is cultivated. Also, in this employment, his knowledge of numbers and counting must be called again into test.

One distinct advantage derived from this occupation lies in the fact that both hands are equally cultivated in skillful action, since both are equally employed and exercised.

The work of Mat Weaving is, from its nature, one of great resource, and the plan appended is submitted merely as an outline or series of hints for guidance in a field of labor for little hands; a field so large that it can be limited only by the length of time or degree of attention which the instructor may judge wise to bestow upon it. Yet, of this he may be certain, that large expenditure of both will be well balanced by the full returns.

1. Object Lessons upon the Materials used.

2. Weaving mats using two colors and strips of equal width in both mat and fringe, by the following formulas (Figs. 215 to 226, page 137), viz.:—

   (*a.*) 1 over, 1 under, 1 over, 1 under, etc., every odd row.

      1 under, 1 over, 1 under, 1 over, etc., every even row.

   (*b.*) 1 over, 2 under, etc., ⎫ in alternate rows as before.
      1 under, 2 over, etc., ⎭

MAT WEAVING. 135

Fig 211.

Fig 212.

Fig 213.

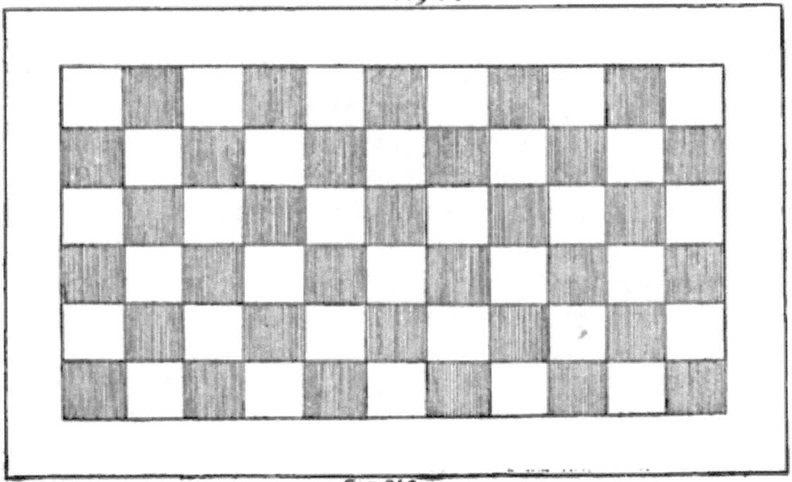

Fig 214.

(c.) 2 over, 2 under, etc., } in alternate rows.
2 under, 2 over, etc., }

(d.) 1 over, 3 under, etc., } in alternate rows.
1 under, 3 over, etc., }

(e.) 2 over, 3 under, etc., } in alternate rows.
2 under, 3 over, etc., }

(f.) 3 over, 3 under, etc., } in alternate rows.
3 under, 3 over, etc., }

(g.) 1 over, 4 under, etc., } in alternate rows.
1 under, 4 over, etc., }

(h.) 2 over, 4 under, etc., } in alternate rows.
2 under, 4 over, etc., }

(i.) 3 over, 4 under, etc., } in alternate rows.
3 under, 4 over, etc., }

(j.) 4 over, 4 under, etc., } in alternate rows.
4 under, 4 over, etc., }

3. Continuing Mat Weaving Lessons by making combinations of any desired numbers, as far as practicable, using two colors.

4. Combining, in the same mat, any two or more of the above formulas, using two colors.

5. Combining any of the above formulas, using any desired number of colors.

6. Weaving easy patterns or figures in the mat. (Figs. 227-237, page 139.)

7. Weaving intricate patterns.

8. Learning to finish each mat by neatly gumming the ends of the fringes to the margin of the mat on the wrong side.

9. Inventive work in Mat Weaving.

# MAT WEAVING. 137

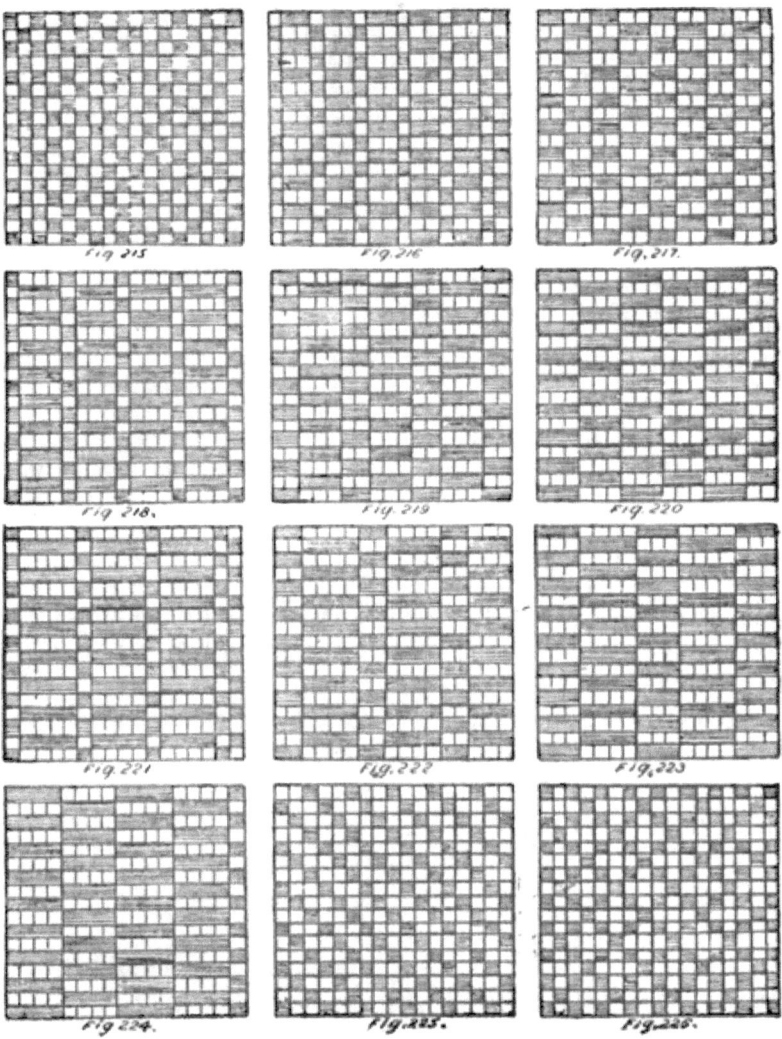

## WRITING.

In the second grade, the work of writing words and sentences upon slate and blackboard is continued. More advanced Tracing and Copy Books are introduced for pen-and-ink writing.

## DRAWING.

The work of the second grade, in drawing, is continued, with much work in invention and design.

## GYMNASTICS.

Free-hand exercises, musical and marching exercises are given, and all previous work reviewed with variations.

[NOTE.—The remarks made under Writing, Drawing, and Gymnastics in Part III., Chap. I., should be referred to.]

MAT WEAVING. 139

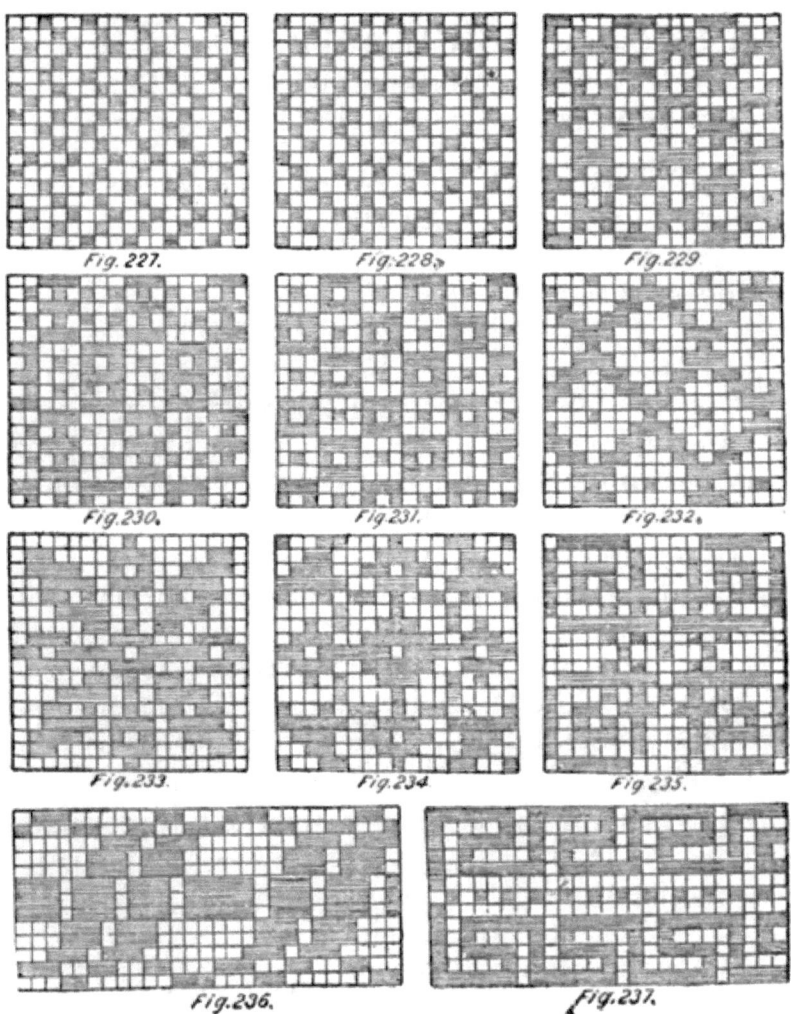

Fig. 227.   Fig. 228.   Fig. 229.
Fig. 230.   Fig. 231.   Fig. 232.
Fig. 233.   Fig. 234.   Fig. 235.
Fig. 236.   Fig. 237.

## Chapter IV.

## SUGGESTIONS, LESSONS, AND METHODS OF INSTRUCTION IN MANUAL ARTS.

### THE PRIMARY SCHOOL—FOURTH GRADE.

1. THE PUPILS. These are usually about eight or nine years old, though many children reach this grade at an earlier age.

2. THE LENGTH OF LESSONS AND AMOUNT OF WORK. As before suggested, daily lessons may be given, or the work may be presented quite profitably upon alternate days. To accomplish good results each lesson should occupy at least twenty minutes.

3. THE STUDIES AND OCCUPATIONS. These are given in detail in Appendix, Chapter II. They comprise Language, Numbers, Sciences (Objective), Manual Training.

THE MANUAL ARTS. The occupations appropriate for this grade are enumerated below,—

Slat Plaiting (advanced).
Crocheting (Chain Stitch).
Paper Folding (advanced).
Perforated Cardboard Embroidery (advanced).
Penmanship.
Free-hand Drawing.
Gymnastics.

## SLAT PLAITING (Advanced).

THE MATERIALS. For this occupation the necessary material has been described under the head of Slat Plaiting (Chapter III., Part Third). In the advanced work these materials may be used with greater freedom and in larger numbers. By way of ornament, inexpensive ribbons and worsteds of various colors may be called into requisition.

---

### LESSONS IN SLAT PLAITING (*Advanced*).

The several occupations in which the child has engaged during the time which he has spent in the Primary Grade, constituting usually the first three years of his school life, have by no means been completed. As each separate employment has been laid aside, it has not been with an absolute sense of relinquishment. The inexhaustible nature of the field of Manual Labor is best proven to the learner by his own experience. A work well done is sure to augment the desire to do it still better.

While in the Primary Grade the fundamental object underlying the beginner's work of making real or simulated forms of use and beauty, was to make him skillful in manipulating and to cultivate his ideas of form, size, color, distance, and number, the more advanced work of the Second Grade still constantly cultivates these ideas, and at the same time the pupil is increasing his capacity to construct things which are really useful. He is becoming more and more skillful in the use of tools and implements.

In reviewing former work he makes better the same forms that he at first made more or less imperfectly.

In the special work of advanced Slat Plaiting now considered, many pretty and useful things may be made. The pupil now may use slats of many colors and lengths. He may construct by instruction, or by his own invention, especially drawing largely upon the latter, many ornamental forms, such as picture frames, card and letter receivers, baskets, mats, etc. He can beautify his work and give its form permanency by means of bows of ribbon or cross stitches of silk or worsted.

He has reached the age when he should learn the names of the geometrical figures, whose construction is especially favored by this employment.

His desire is easily stimulated to make useful articles for mother or sister at home, and although the gift be very rude and simple, the consciousness of ability to create, and the sensation of pleasure awakened by doing for others, are to every young person, if properly directed, priceless acquisitions.

The advantages gained by systematic and frequent reviews cannot be overvalued.

## CROCHETING CHAIN STITCH.

THE MATERIALS. These are the crochet needle, which may be of bone, ivory, wood, steel, or rubber; and the thread of yarn, worsted, cotton, or linen.

They may be procured by the quantity, of the usual dealers at reasonable expense.

### LESSONS IN CROCHETING CHAIN STITCH.

This occupation is introduced, like all previous ones, by Object Lessons on the materials.

Next, the pupil is instructed in the proper method of holding the needle and the yarn. (It is prefer-

# CROCHETING CHAIN STITCH.

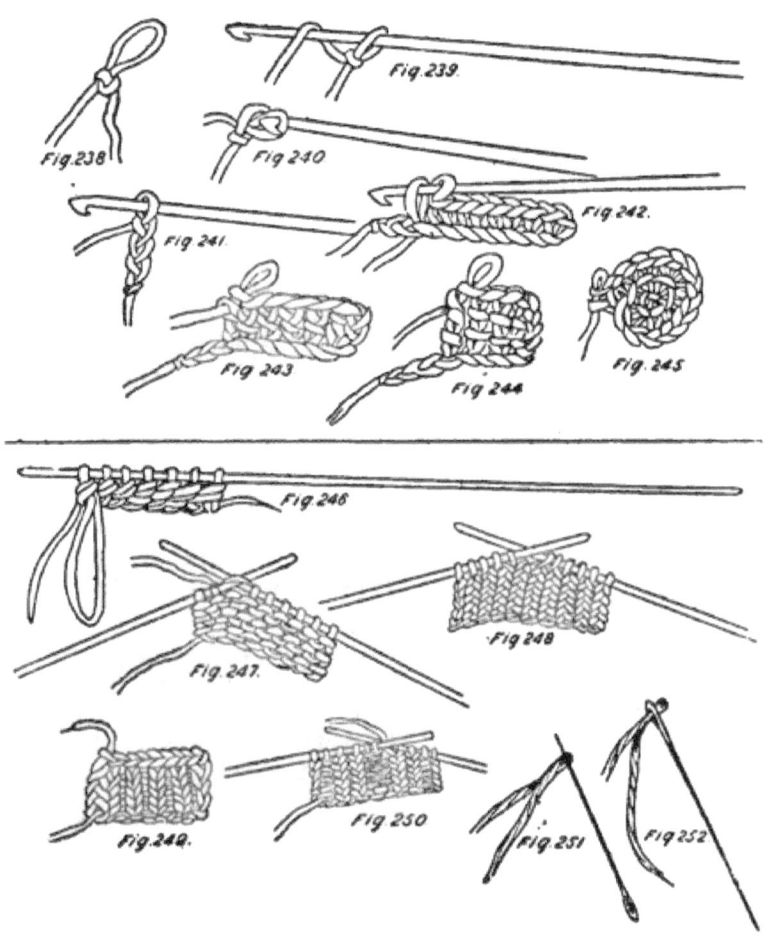

able to use yarn or coarse worsted for beginning lessons in crocheting.) A child who holds his needle and work awkwardly must be assisted and reminded persistently. He can attain grace by practice. Ungraceful movements in any occupation should be striven against with unceasing effort. Verbal instruction in crocheting must, of necessity, be strongly supplemented by the example of the instructor. Much individual attention, in the beginning, will aid pupils who do not crochet "naturally," to form good habits.

The first easy lesson on the simple Chain Stitch is presented as follows:—

The class being supplied with a crochet-hook and ball of yarn for each pupil, all are instructed to make a loop or slipping-knot in the end of the yarn. (Fig. 238, page 143.)

Next, to wind the yarn once around the fourth finger of the left hand; to let it pass under the third and second fingers and over the first, and to hold the loop between the thumb and the first finger.

Next, to take the crochet-needle in the right hand, in the same manner as a pencil is held; to put the hook through the loop; to put the hook under the yarn that passes over the first finger, and to draw it through the loop; again to put the hook under the yarn and draw it through the new loop; to continue repeating this process. (Figs. 239, 240, 241.)

The pupil must be taught to hold the needle so that the hook will always be down, or nearly so; to hold the yarn loosely and to make the loops uniform and the chain straight and smooth; not to unwind too much yarn from the ball at one time.

For practice lessons the pupils may crochet chains of different lengths by counting, using different materials. They should also learn how to ravel the work and rewind the yarn neatly upon the ball.

## *PAPER FOLDING (Advanced).*

THE MATERIALS. The description of these materials and their expense, as well as the manner of procuring and preparing them, have been given under the head of Paper Folding, Chapter III., Part First.

**LESSONS IN PAPER FOLDING** (*Advanced*).

From the simple forms of Paper Folding, which constituted the beginning lessons in this employment, the pupil is now prepared to advance to those of a more complicated nature. No caution, before observed, in regard to cleanly and accurate work, can now be omitted, but, on the contrary, all requirements of that nature should be held to still more strict observance.

Each one of these advanced forms of Paper Folding embraces sufficient work for a single lesson. These lessons will be described in order.

### FIRST LESSON.

THE TABLE CLOTH. According to former instructions in Paper Folding, make in the order named, the book, window, shawls, open envelope, closed envelope. Next, unfold the paper; place it upon the desk or table; find the middle horizontal fold. Carefully fold the horizontal edge nearest you, to meet this middle fold; turn the paper half around; fold opposite edge to meet the middle fold, as before, and crease; unfold; fold the two remaining edges to meet the vertical middle fold in like manner. Now fold together the four edges to the middle, and crease the square thus made. Also crease the diagonal fold in each corner square.

We have a table cloth with four hanging corners. (Fig. 253, page 149.)

Rule to remember: Fold *from you* and *by opposites*.

### SECOND LESSON.

THE TABLE CLOTH GROUND FORM. Make the table cloth just described. Crease the large middle square. Crease the four corner squares to the center. (Fig. 254.)

The form thus made is the fundamental folding for many beautiful designs.

### THIRD LESSON.

THE WINDMILL. Make the table cloth form just described.

Turn toward you, successively, each side of the figure, and fold out each one of the corner squares in the plane of the whole form, creasing the diagonal fold. Turn all the points toward the right, or, all toward the left hand. (Fig. 255.)

### FOURTH LESSON.

THE WATER WHEEL. Make the Windmill. Fold back each of the projecting triangular folds upon its base, creasing. (Fig. 256.)

### FIFTH LESSON.

THE VASE. Make the Windmill. Crease the middle diagonal fold making two points extend upward. The remaining two points fold downward to form the base of the figure. (Fig. 257, page 149.)

### SIXTH LESSON.

THE SAIL BOAT. Make the Vase. Fold one of the triangular forms at the base, over upon the other, and crease. (Fig. 258.)

### SEVENTH LESSON.

THE DOUBLE BOAT. Make the Vase. Crease the horizontal instead of the diagonal fold. This will place two little boats side by side. (Fig. 259.)

### EIGHTH LESSON.

BOAT WITH FISH BOX. Make the Double Boat. Turn back and crease the small triangles which form the bow and the stern of one of the boats. (Fig. 260.)

### NINTH LESSON.

THE OLD HEN. Make the Vase. Fold back one of the small triangles at the base. The one remaining is the old hen's wing. Skillfully turn wrong side out one of the remaining points at the top of the vase. This makes the hen's head and bill. (Fig. 261.)

### TENTH LESSON.

THE SWIMMING DUCK. Make the old Hen. Fold back the small projecting triangle which formed the hen's wing. (Fig. 261, page 149.)

---

### ELEVENTH LESSON.

THE DOUBLE TABLE CLOTH FORM. This form, like the single table cloth form, is used as the foundation for all of the forms described under that head, and for many whose details need not be given here. It is made as follows: Make the closed envelope. Now crease the folds which form this square. From the square thus made placed upon the desk with the triangular flaps uppermost, make a table cloth. From this make a table cloth form as described in Lesson Second. The result will be a reduced size of the single form represented.

---

It goes without saying that the child takes much delight in constructing these interesting forms of life. He can make a fleet of boats of different colors. Cutting several of the squares of paper, each into four smaller squares, he can make a whole brood of chicks for the old hen, or of ducklings for the old duck. Much time should be given for amusement in this occupation. The teacher should have at his ready command many pleasant little stories. These and numerous small games may be introduced frequently when appropriate to the regular work of the lesson in hand, and serve to vary the routine of class work very pleasantly.

PAPER FOLDING (Advanced). 149

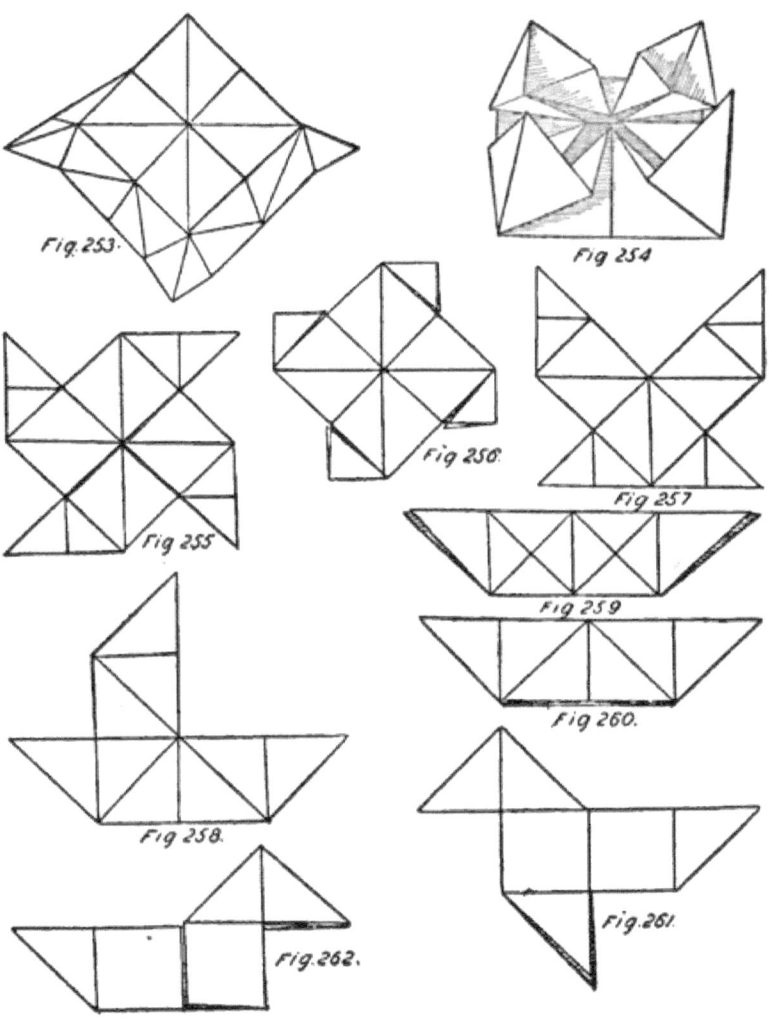

## PERFORATED CARDBOARD EMBROIDERY
*(Advanced).*

THE MATERIALS. In addition to the materials enumerated and described under the head of Perforated Cardboard Embroidery (Chapter III., Part Third), use can be made of ribbons, silks, and a great variety of worsteds, in the advanced work now presented. But no great expenditure is necessary, as the economical use of a few pretty materials will insure the result to be quite satisfactory. Any tendency to useless expenditure should be promptly checked.

### LESSONS IN ADVANCED PERFORATED CARDBOARD EMBROIDERY.

GENERAL OUTLINE OF WORK. Beside following out the detailed scheme of work laid down in the Primary course, the pupil, doubtless, has made since, considerable advancement in this employment at the various opportunities afforded for reviewing former work. Now, taking it up at its second recurrence in the Manual Curriculum, he may be given large license for the creation of objects in cardboard, many of which are fanciful and not a few, if well made, useful.

That a definite plan of labor may not be wholly unobserved, however, the following may be found of use, subject to such variations as may seem wise to the instructor in its adoption and observance:—

1. Review of former work in Cardboard Embroidery.

2. Embroidering simple borders, using one kind of stitch, and employing one, two, three, and, finally, any number of colors of worsted.

3. Embroidering simple borders using two, three, finally, any number of kinds of stitches, and employing but one color of worsted.

4. Making simple borders, using any number of kinds of stitches, and employing any number of colors desired. (Page 128.)

5. Employing any of the above described combinations in embroidering simple symmetrical designs, not borders.

6. Embroidering intricate designs and borders in the same manner.

7. Embroidering letters, figures, and symbols, in various styles.

8. Embroidering forms of life, viz., animals, flowers, fruit, etc.

9. Using various kinds and colors of cardboard.

10. Embroidering objects of use. When embroidered, learning to finish the object neatly by lining, binding, and ornamenting with tassels or balls of worsted, bows of ribbon, etc. (This perhaps will come under the head of review work, when the child has had some experience in sewing.) The list of these pretty objects is almost endless. It includes the book-mark, card-receiver, hair-receiver, letter-case, handkerchief-case, spectacle-case, and a great variety of other articles.

11. Invention in Cardboard Embroidery.

## PENMANSHIP.

The work of the previous years is reviewed with variations, the pupils writing both with and without copies. More advanced Tracing and Copy Books are presented.

## DRAWING.

The previous work is reviewed, and much free-hand and inventive work required on the slate and the blackboard.

---

## GYMNASTICS.

Free-hand exercises and marching are systematically given.

[NOTE.—The remarks made under Writing, Drawing, and Gymnastics in Part III., Chap. I., should be referred to here.]

# Chapter V.

## *SUGGESTIONS, LESSONS, AND METHODS OF INSTRUCTION IN MANUAL TRAINING.*

#### THE PRIMARY SCHOOL—FIFTH GRADE.

1. THE PUPILS. These children, entering upon their fifth year at school, will range in age from eight to nine or ten years.

2. LENGTH OF LESSONS AND AMOUNT OF WORK. Directions given for Fourth Grade will be found applicable for this one.

3. STUDIES AND OCCUPATIONS. Appendix, Chapter II., gives in detail an outline for this work. They cover four subjects, viz., Language, Numbers, Sciences (Objective), Manual Arts.

4. THE MANUAL ARTS. For this grade the following occupations have been found admirably adapted,—

>Sewing Over-and-Over.
>Crocheting.
>Paper Folding and Mounting.
>Penmanship.
>Free-hand and Inventive Drawing.
>Gymnastics.

## SEWING OVER-AND-OVER.

THE MATERIALS. The child now takes the implements and materials in his hand for the first time, for the purpose of what seems to him "real work," since it is his first introduction into the mysteries of "plain sewing." The chief attraction of this employment, lies, perhaps, in the fact that the pupil approaches more nearly than he has at any former time, the sphere of "grown-up" work. To nearly all child-natures this is a common ambition.

All ordinary implements and materials necessary for common sewing are employed. The cloth used for the beginning lessons is unbleached muslin, cut into pieces about five inches square. A very few yards will do service for a long time, as the stitches may be removed and the same piece re-used many times. As the work proceeds, cloth of different kinds and textures may be employed.

---

### LESSONS IN SEWING OVER-AND-OVER.

Object Lessons on all of the implements and materials used will be productive of interest and profit. When these have been completed and at the proper time, the class being in readiness for sewing, to each member should be given a needle of proper size, a spool of thread of proper number, a thimble, an emery-ball, a small piece of bees-wax, a pair of scissors, and two of the pieces of cloth above described. All of these should be neatly arranged for each pupil, in a small work-basket or box, labeled with his name.

He should then be instructed as follows:—

*First.* To smooth neatly the pieces of cloth and to trim the edges if necessary.

*Second.* To turn down narrowly and crease a corresponding edge of each piece.

*Third.* To baste neatly each of the folds thus turned.

*Fourth.* To place the turned edges of the two pieces exactly together, with the turned over edges either outside or inside as directed, and to baste the pieces together firmly, in this position.

*Fifth.* To commence sewing at the right hand side of the work, by drawing the needle and thread through the cloth, very near to the folded edge, concealing the knot.

*Sixth.* To thrust the needle, toward the sewer, through both folds, very near the edge, and to continue " over-and-over," toward the left hand. (Fig. 295, page 168.)

*Seventh.* To make small, evenly placed stitches, and to guard against taking them too deeply, or drawing them too tightly. Not to use a very long thread.

*Eighth.* When some progress has been made and longer seams are given, to learn how to pin the work to knee or waist to hold it firmly while sewing.

*Ninth.* To learn to fasten the thread neatly when the seam is finished.

*Tenth.* To learn to smooth down and press open a finished seam.

*Eleventh.* To learn to take out work, without injuring the fabric, removing the stitches carefully.

*Twelfth.* At all times to use the scissors, not the teeth, in cutting the thread.

*Thirteenth.* To learn the use of the wax when the thread has a tendency to tangle, and how to use the emery-ball when the dull needle "squeaks" in the perspiring little fingers.

*Fourteenth.* Never to enter the sewing class with untidy hands. To make every possible effort to keep the work clean.

*Fifteenth.* To place all materials in proper order and place at the close of every lesson.

The pupil should not be allowed to change from this occupation to any other until he has attained in it a fair degree of proficiency. Though his patience and fingers tire, he should still be encouraged by the instructor to persevere in view of a reward of good results, and thus the value of the lesson will be doubled.

## CROCHETING.

THE MATERIALS. Under the heading, Crocheting Chain Stitch (Chapter III., Part Fourth), will be found a description of the required materials for the present employment. They will require no variation or addition.

### LESSONS IN CROCHETING.

SIMPLE STITCHES. The chain stitch forms the foundation or beginning of all simple forms of crochet work. After reviewing that stitch the various kinds of easy stitches and simple forms of crocheting may be presented in successive lessons, as follows:—

### FIRST LESSON.

SINGLE OR PEARL STITCH. Make a chain of twenty, thirty, or any given number of stitches.

Turn the work.

Thrust needle through upper thread of next to last stitch made, and under the worsted, and draw through. Put worsted over and draw through the two stitches on the needle.

Thrust needle through upper thread of next stitch of chain.

Repeat as before. (Fig. 242, page 143.)

When a row has been completed, turn the work, and crochet across in the opposite direction, taking up every stitch as before. At turning crochet one chain. To make a right and a wrong side to the work, break the thread at the completion of each row and fasten neatly and securely by drawing the end through the final stitch or loop. Then attach the thread nicely at the opposite end, and work as in first row. To vary this stitch, take up as desired,—the upper thread only; the lower thread only; or both threads of the stitches in the preceding row.

### SECOND LESSON.

THE DOUBLE OR SLIPPER STITCH. Make a chain.

Turn work and crochet two chain.

Put thread over and thrust needle through upper thread of third from last stitch of chain, and draw thread through.

Put thread over and draw through two of the three stitches on the needle.

Put thread over and draw through the two remaining stitches on the needle.

Repeat, taking up every stitch. (Fig. 243, page 143).

Observe the same changes in taking up stitches in succeeding rows, as given in directions for Pearl Stitch.

Vary the exercises in this stitch by crocheting one chain stitch between each of the stitches and omitting a corresponding stitch in taking up the stitches of the preceding row.

Make two chain between in the same manner.

Make three between in the same manner.

---

### THIRD LESSON.

TRIPLE STITCH. Make chain.

Turn work, crocheting three chain.

Put thread over twice and thrust the needle through third from last stitch of chain, and draw thread through.

Put thread over and draw through two of the four stitches or loops on the needle.

Put thread over again and draw through two of the three stitches remaining on the needle.

Put thread over again and draw through the two stitches remaining on the needle. (Fig. 244, page 143.)

Repeat.

Use the same variations in this stitch as those described for previous stitches.

---

### LESSONS FOR PRACTICE IN CROCHETING.

Much practice must be given in crocheting, chiefly at first for the attainment of ease and grace in holding and managing work and needle. There is much connected with crochet work that cannot readily be imparted as instruction by one person to another. One who crochets "naturally" finds frequent opportunity, and indeed necessity for the use of judgment regardless of rules. The learner will understand this better as he becomes more and more familiar with the work. Too strict an observance of rules and formulas is often a hindrance to the accomplishment of good crochet work. Miscellaneous, careless, or hasty work should, of course, be condemned, but good judgment and original method in crocheting are of the first necessity.

Much practice in easy work, leading gradually to more difficult tasks, will accomplish results for the pupil which no amount of instruction would enable him to attain.

The plan below suggested covers much ground for practice.

1. Winding worsted into soft, smooth, shapely balls.

2. Crocheting chain slowly, counting stitches as made; striving to avoid awkwardness of movement, and endeavoring to hold work and needle easily.

3. Making chain more and more rapidly, counting stitches.

4. Crocheting chain of any given length and fastening the ends together, forming a ring.

5. Taking up each stitch of the ring thus made, crocheting "round and round" either in pearl, double, or triple stitch. Fastening at the end of each round and crocheting one, two, or three chain, as necessary, before beginning new row.

6. Learning to narrow and widen by omitting stitches in the former case and supplying them in the latter.

7. Crocheting a little mat employing pearl, double, or triple stitch. Practicing to make the mat very flat and smooth. (Fig. 245, page 143.)

8. Making simple edgings by combining the various stitches learned.

## PAPER FOLDING AND MOUNTING.

Under the headings Paper Folding (Chapter III., Part First), and Scrap-Book Making (Chapter III., Part Second), the materials for the occupation for this grade will be found fully described.

---

**LESSONS IN PAPER FOLDING AND MOUNTING.**

To the occupation of Paper Folding the pupil cannot fail to revert with fresh interest at its every presentation. It is doubtful if in any other one the revelation of new wonders produced by his hands is so constant and so great. While much in it is old, and he must draw frequently upon his stock of previous knowledge, there are still open to his advance unnumbered avenues yet unexplored. A few hints, and those of the briefest, are all that he will require for guidance.

The instructor, therefore, will find his tact and prudence called into exercise more for the purpose of reserving instruction than for imparting it. The unconscious injustice often done to the pupil by telling too much, and so robbing him of the privilege of finding out for himself, is proven by his weakness when his powers are called to any test. In every branch of instruction this is too often forcibly illustrated, and the subject now under discussion affords so excellent a field for the exercise of original method and natural ingenuity that the danger of crippling the grand power of invention in the beginning of its activity, should be especially guarded against.

# PAPER FOLDING AND MOUNTING. 161

Fig 263. Fig 264. Fig 265. Fig 266.
Fig 267. Fig 268. Fig 269. Fig 270.
Fig 271. Fig 272. Fig 273. Fig 274.
Fig 275. Fig 276. Fig 277. Fig 278.
Fig. 279. Fig. 280. Fig 281. Fig. 282.

11

The instructions laid down in the chapter on Scrap-Book Making, in regard to the pasting and mounting of pictures, will be found to be directly applicable to the mounting of forms produced by Paper Folding.

A brief plan for work, covering an unlimited number of lessons, is suggested as follows:—

1. Review of former work in Paper Folding.

2. Making simple and complicated forms from the closed envelope or simple form as a foundation.

3. Making simple and complicated forms from the single table cloth ground form.

4. Making simple and complicated forms from the double table cloth ground form. (Figs. 263 to 282, page 161.)

5. Cutting one of the squares of paper into four small squares; making small forms from these, in the various ways described above.

6. Mounting any of the forms made by Paper Folding in various positions, singly.

7. Mounting any of the forms made, in groups of two, three, four, or any desired number, combining different colors. (Figs. 283–294, pages 163, 165.)

8. Especial opportunity and much time for invention in Folding and Mounting.

## PENMANSHIP.

The previous work on slate and blackboard is reviewed and continued. Paragraphs from the Reader are written daily in connection with regular lessons. More advanced Tracing and Copy Books are introduced.

## PAPER FOLDING AND MOUNTING.    163

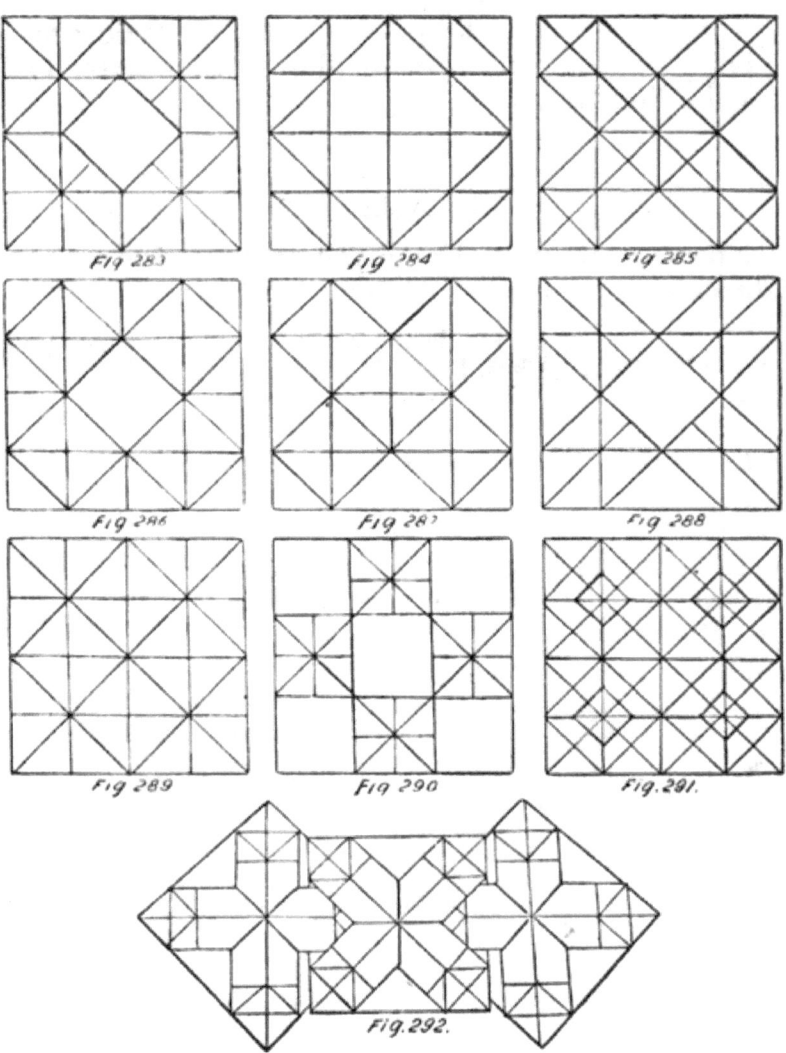

Fig 283.   Fig 284.   Fig 285.
Fig 286.   Fig 287.   Fig 288.
Fig 289.   Fig 290.   Fig. 291.
Fig. 292.

## DRAWING.

The work for this grade consists of review and advancement upon the same plan as that described for the Fourth Grade.

## GYMNASTICS.

New exercises in Free-hand Gymnastics and in Marching added to those given as in previous grades.

[NOTE.—The remarks made under Writing, Drawing, and Gymnastics in Part III., Chap. I., should be referred to here.]

PAPER FOLDING AND MOUNTING. 165

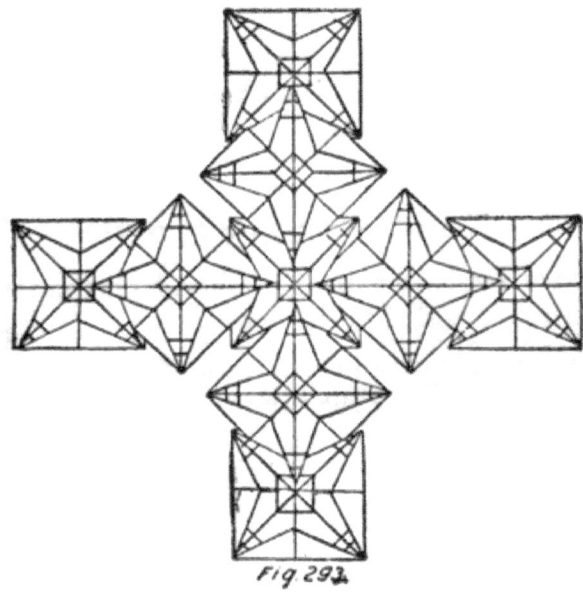

Fig. 293.

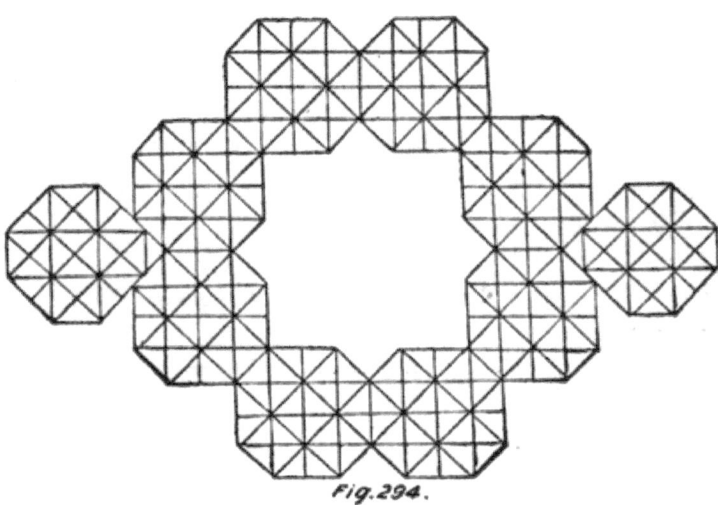

Fig. 294.

# Chapter VI.

## SUGGESTIONS, LESSONS, AND METHODS OF INSTRUCTION IN MANUAL TRAINING.

### THE PRIMARY SCHOOL—SIXTH GRADE.

1. THE PUPILS. These are nine or ten years of age as a rule, though some complete the fifth year's work earlier than this and some take a longer time to do it.

2. THE LENGTH OF LESSONS AND AMOUNT OF WORK. The time at the teacher's disposal, the resources at his command, and the convenience afforded him, must establish these *data*. General programs must be made to conform to individual needs. If more time and more frequent lessons can now be allowed, the result will be greater. An excess of thirty minutes for each day in any given lesson is not advised.

3. THE STUDIES AND OCCUPATION. Reference to Appendix, Chapter II., will give in outline the work appropriate for this grade. The subjects covered are: Language, Numbers, Sciences (Objective), Manual Arts.

4. THE MANUAL ARTS. These are,—

| | |
|---|---|
| Hemming. | Penmanship. |
| Pease Work. | Drawing. |
| Knitting. | Gymnastics. |
| Paper Flower Making. | Reviews. |

## HEMMING.

THE MATERIALS. The materials and implements necessary for this occupation are the same as those required for sewing "over-and-over." The object lessons there given on these materials and implements may now be reviewed with profit.

---

### LESSONS IN HEMMING.

*First.* The pupil is taught to turn a raw edge of the cloth narrowly toward the worker; to crease the fold thus made, and to baste it neatly.

*Second.* To turn this folded edge toward him in the same manner, thus completing the folding of the hem. To baste this fold as before, taking great care to preserve the uniform width of the hem. (Fig. 296, page 168.)

*Third.* To fold and baste hems of various widths.

*Fourth.* To draw the needle and thread through the folded edge at the left hand, concealing the knot beneath the folded edge; to hem the folded edge down neatly upon the body of the cloth, striving to make small stitches of uniform length and slant, to take up as little of the cloth as possible in making the stitches. Not to use a very long thread nor draw it too tightly. (Fig. 296.)

*Fifth.* To fasten the thread neatly at the end of the hem and to remove the basting thread.

*Sixth.* To observe unfailingly all of the general directions in regard to workmanlike sewing, given under the head of "Over-and-Over."

*Seventh.* For practice to make some article or garment in miniature as, a bag, a sheet, a pillow-case, employing the two kinds of stitches learned in sewing.

168  INDUSTRIAL EDUCATION.

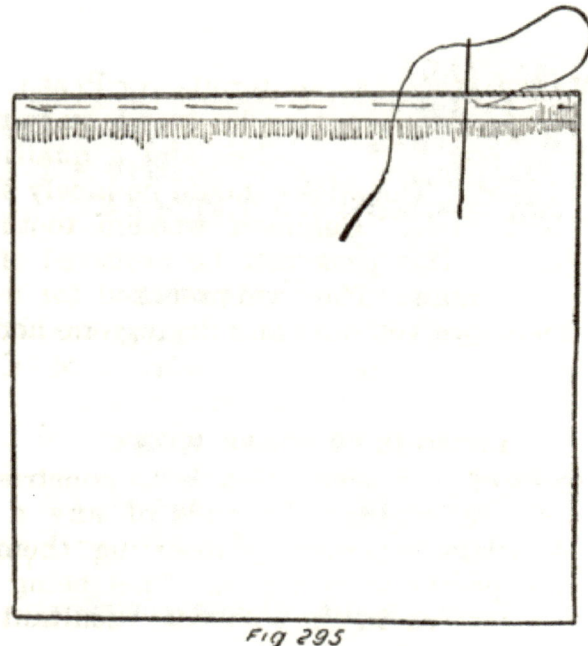

Fig 295.

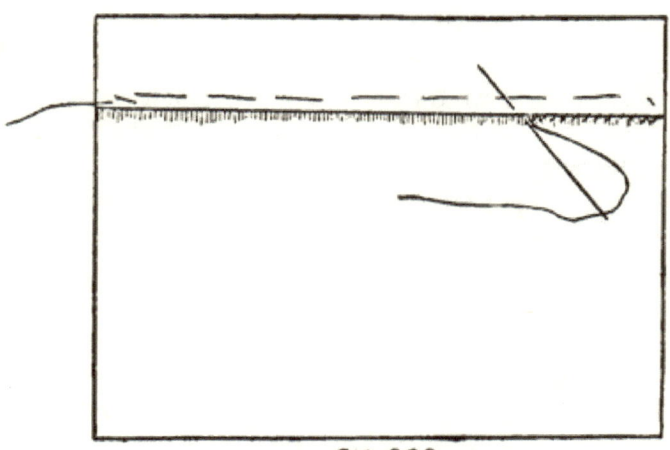

Fig 296.

## PEASE WORK.

THE MATERIALS. The materials for Pease Work consist of small sticks of pine wood, varying in length from one to five inches, and a quantity of good sized peas. The sticks should be nicely sharpened at both ends. Common wooden toothpicks may be used. The peas can be procured of any grocer or seedsman. They are prepared for use by soaking them twelve hours and drying one hour.

### LESSONS IN PEASE WORK.

The object of this occupation is to construct various forms, by joining the ends of any desired number of sticks together by inserting them into peas at the points of juncture. This being done while the peas are in the soft state resultant from soaking them, when they become dry and hard any figure thus made will be a permanency.

The figures may lie in the plane upon which they are originated, or they may be constructed upon it. The pupil is privileged to draw largely upon the resources already developed in the occupations of Block Building, Stick Laying, and Slat Plaiting, since the underlying principles of all of these employments are closely related. Herein lies the special charm of Pease Work, that it embodies an old idea, or set of ideas, in forms so new and pleasing and by means of an occupation so diverting, that the worker feels no irksome consciousness of too frequent reiteration.

On account of the liability to split the pea, the pupil should learn to insert the stick into the side of a hemisphere, not too far, and not into the line or space which divides the pea into two parts.

When a piece of Pease Work is finished, it should be placed where it can remain undisturbed until thoroughly dry.

For practical application of these hints the following plan is suggested for a series of lessons:—

1. Review of former lessons in symmetrical plane figures made by Stick Laying.

2. Object Lesson: subject, The Pea.

3. Practice with a single stick and pea, learning to properly insert the former into the latter.

4. Using in successive lessons one, two, three, and four sticks of equal length, in combination with one, two, three, and four peas, advancing from simple to complex forms. (Figs. 297 to 303, page 171.)

5. Same as above, employing sticks of different lengths.

6. Combining any number of peas and sticks in plane figures. (Figs. 304 to 314, pages 171, 172.) Drawing these figures upon the board or slate.

7. Same as above in elevated forms. (Figs. 315 to 324, pages 172, 173.)

8. Invention in Pease Work.

## KNITTING.

THE MATERIALS. The final occupation in the Industrial curriculum for the Primary and Second Grades is now reached. The pupil, having learned the rudiments of sewing and crocheting, may take with equal profit the initiatory steps in the art of Knitting.

PEASE WORK.

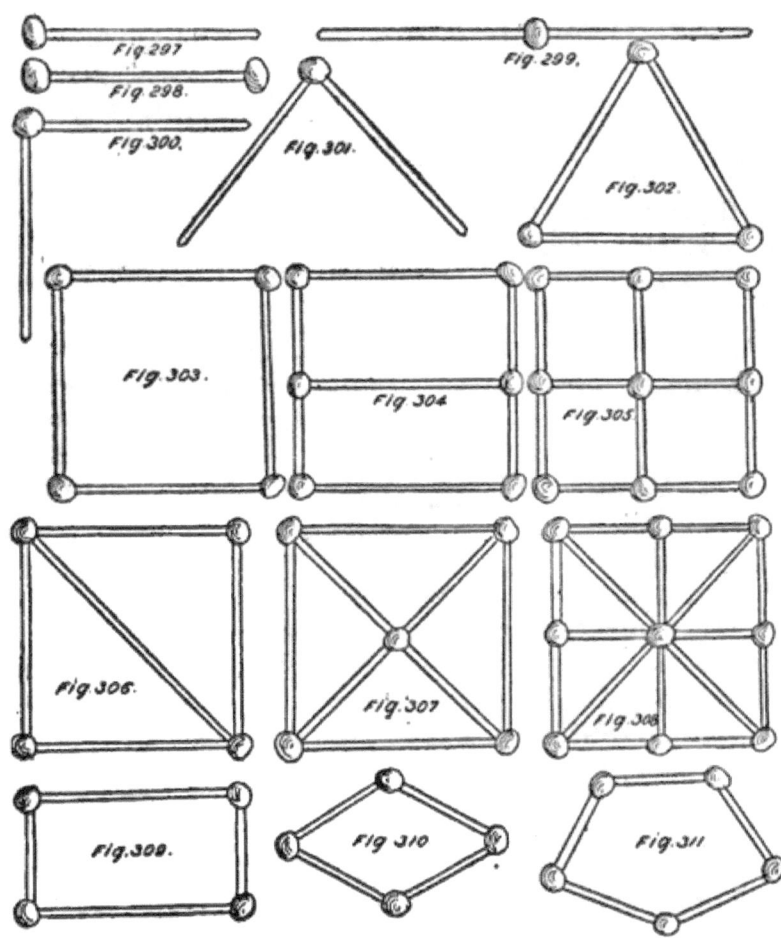

172                INDUSTRIAL EDUCATION.

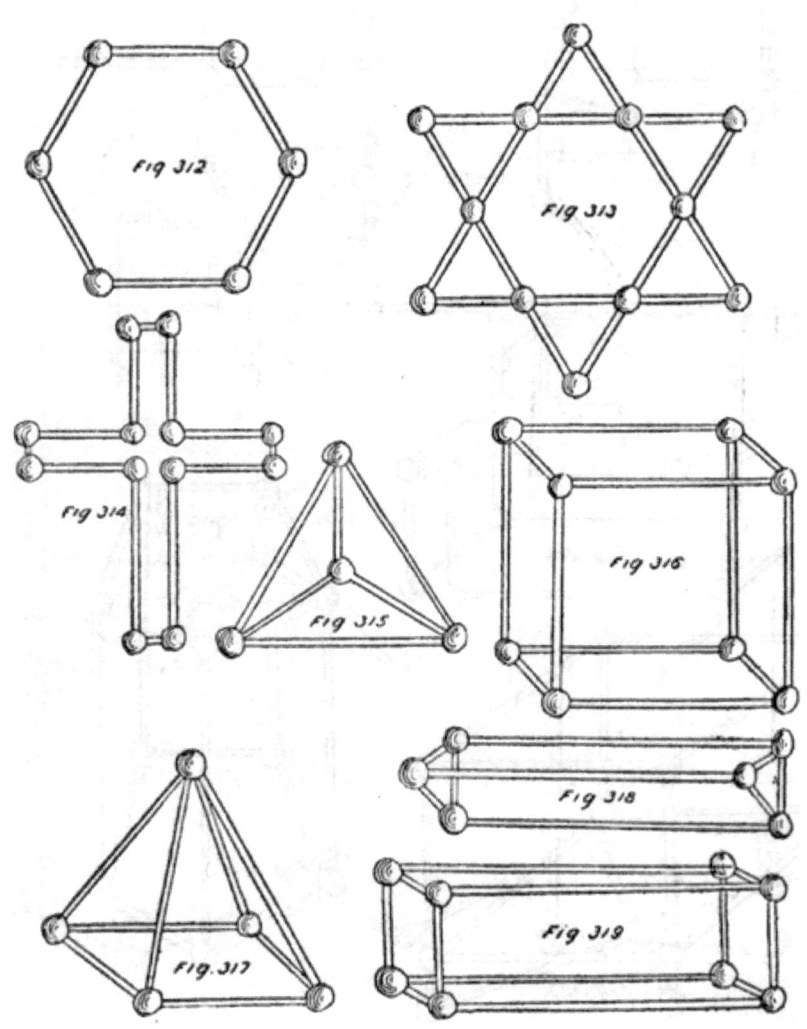

PEASE WORK.    173

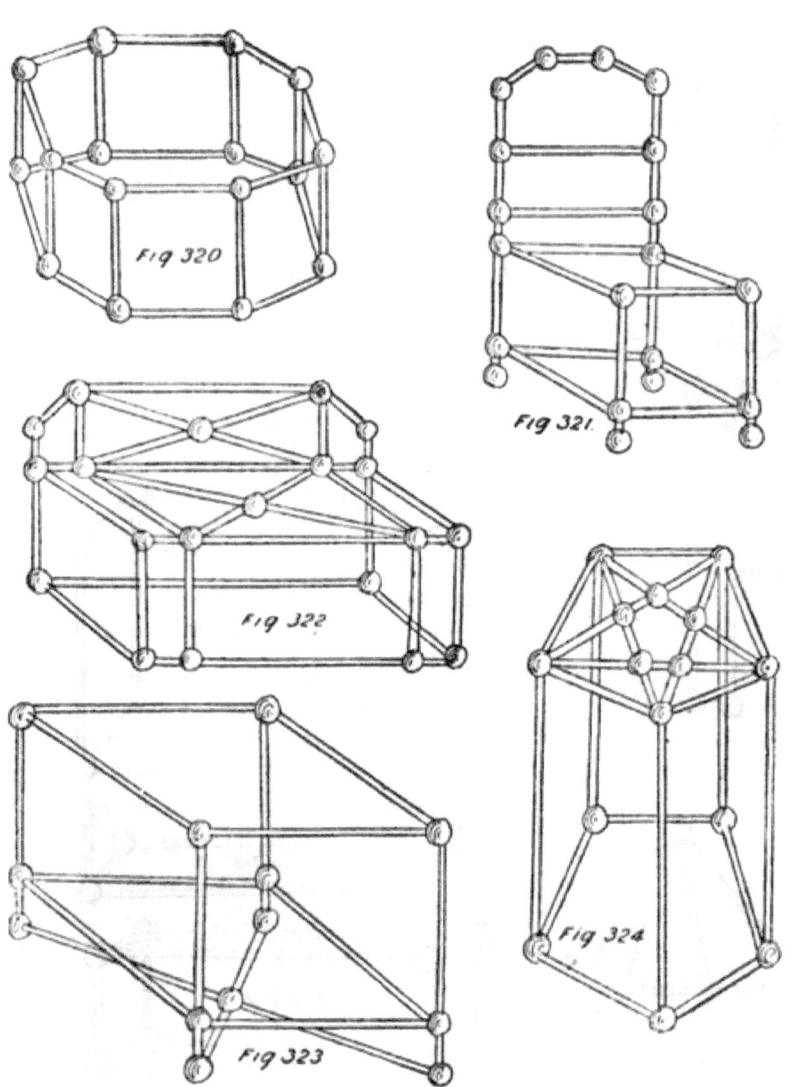

Fig 320

Fig 321.

Fig 322

Fig 323

Fig 324

The necessary materials consist of a quantity of common knitting yarn, and for each child a pair of good steel knitting-needles of size to suit the yarn. These are procured of the dealers as described under Paper Embroidery, Crocheting, etc.

---

**FIRST LESSON.**

CASTING ON STITCHES. Present to each child one needle and a ball of yarn, and instruct as follows:—

Double the end of the yarn one-half yard. Pass the single yarn over the first finger of the right hand.

Hold the doubled part of the yarn in the left hand passing over the thumb.

Knitting-needle in right hand.

Put the first finger of the left hand through the yarn which is around the thumb, drawing it to the finger.

Push the needle under the yarn which is on the finger of the left hand, and put the yarn which is in the right hand over the needle.

Draw the top of the needle toward you, and the yarn off the finger.

Tighten the yarn and the first stitch is made.

Repeat until the doubled yarn is used. (Fig. 246, page 143.)

The teacher should illustrate these instructions by performing the work with the class, repeating, and giving individual aid until all have accomplished it. Then they should practice until they are able to perform it rapidly and easily.

### SECOND LESSON.

PLAIN KNITTING WITH TWO NEEDLES. Cast on twenty stitches (or any given number).

Change the needle upon which are the stitches to the left hand.

Take a needle in the right hand.

With this needle take off, one at a time, the stitches from the left hand needle, in the same manner as you did from your finger, in "casting on." This time do not tighten the stitches.

When you have removed them all to the right hand needle, change needles and repeat. (Fig. 247, page 143.)

This lesson should receive much practice before a new one is presented.

---

### THIRD LESSON.

LEARNING THE SEAM OR PURL. Cast on twenty stitches.

Knit first row plain.

Slip first stitch off upon your needle by putting the needle through opposite side of the stitch.

Put needle through next stitch in same way.

Put yarn around the needle toward you, and to draw the stitch off, push the needle from you.

Take the twenty stitches off in same manner.

Knit back plain.

Purl next row.

Repeat, knitting alternate rows, plain and purl.

This knitting will have a right and a wrong side. (Fig. 248, page 143.)

## LESSONS FOR PRACTICE IN KNITTING.

1. Knitting a strip alternating a plain and a purl stitch. (Fig. 249.) In successive lessons alternating, as follows:—

>Two plain, one purl, etc.
>Two plain, two purl, etc. (Fig. 250.)
>Three plain, one purl, etc.
>Three plain, three purl, etc.

2. Learning to "bind off" by slipping the stitches one over another until but one stitch is left upon the needle.

3. Learning to knit a simple edging from directions written upon the blackboard.

---

## *TISSUE-PAPER FLOWERS.*

**The Materials.** The materials required for paper-flower making consist, first,—of sheets of tissue paper, which can be procured in any color desired from book dealers, at the rate of 2 cts. to 14 cts. per sheet, or 40 cts. to $2.75 per quire; second,—single and branched leaves and stems from 1 ct. to 3 cts. each, or 10 cts. to 30 cts. per doz.; third,—culots (or centers for roses, daisies, poppies, etc.), at 30 cts. per doz.; fourth,—seed-pods, buds, rubber stems, etc., at 4 cts. to 20 cts. per doz.; fifth,—working tools, consisting of tweezers, goffering-sticks, and other necessary tools, also a bottle of good mucilage. The prices of all of the articles mentioned vary according to kind and quality. A small quantity of material will suffice to make a very good beginning. Additions to the original stock can be made from time to time at trifling expense.

## LESSONS IN PAPER FLOWER MAKING.

INTRODUCTORY HINTS. The work of paper flower making has been received since its introduction with such universal favor that it will probably require no special recommendation to secure it a place among the occupations for manual practice in the school. There are several reasons why its presentation to pupils is appropriate, and, as a natural consequence, of value. First, — delicacy of touch and skillful manipulation are acquired. Second,—color lessons are continued under most favorable circumstances. Third,—the taste is cultivated. Fourth,—valuable lessons from one of Nature's most beautiful and varied pages must necessarily be gained, for as different kinds of flowers are introduced, their similarities and differences are noted, and there are numberless interesting details which careful guidance and instruction are certain to evolve; the parts of a flower are learned, also their arrangement, number, and mode of attachment.

It would hardly be profitable to present this occupation until the pupils had shown, by the work accomplished with other delicate materials, that preparation which is necessary for an intricate work requiring the greatest patience and care.

Before the work of flower making proper is begun there are many preliminary details to learn. The pupils ought to become acquainted with the materials and tools and to learn their names. Some opportunity should be given for practice in the various manipulations by which the parts of a simple flower are prepared, before the pupils advance to the construction of the flower as a whole.

Experience in the previous occupations of paper folding and cutting will enable them to take up quite readily the first simple lessons in flower making. The other processes should follow in the natural order which a need of them will suggest.

### FIRST LESSON.

FOLDING AND CUTTING. Take for practice a piece of common wrapping paper of light texture. Cut a four-inch square. Mark the corners A B C D by oppposites. (Fig. 325, page 183.) Carefully fold A over diagonally upon B and crease sharply. Open the paper and in like manner fold C upon D. Then fold C back to A and D to B. Press the folds of the triangle thus folded and having marked upon it the shape of the petal desired cut through the eight thicknesses, following the line indicated. (Fig. 326.)

Some kinds of flowers may be made by cutting each petal separately and pasting them one by one in place; but it will be readily seen that a simple way of saving time and labor is to cut a set of petals and leave them connected in the center. This is especially convenient in making rows of petals for double flowers, the outer one excepted. A set of flower petals thus united in the center is called a Slip. (Fig. 327.) In the process of making up a flower, putting these sets of petals in place upon the stem is called Slipping.

In a similar manner to that described above, the paper may be folded to make six thicknesses if desired. (Fig. 328.) The method of folding the paper back and forth in a fan-like manner is better than that of folding by doubling and redoubling, since by it more uniform size and shape are secured in cutting the petals. But too great precision should be avoided for the greatest charm in Nature's productions which we are endeavoring to imitate, is often found in their vagaries.

After practice in folding and cutting pieces of wrapping paper, the pupils may cut petals for some simple flower, from tissue paper of the proper color. They may also cut a number of slips and lay them all aside for future use.

### SECOND LESSON.

CRIMPING. Petals of different flowers may be crimped in several ways to produce good imitations of natural forms. The best way to crimp a petal through the middle is to fold it lengthwise over a stiff knitting needle, and crimp by pressing small folds with the left hand. The right hand receives and holds with the thumb and first finger the crimps thus made. (Fig. 329, page 183.) After a little practice this can be done easily and rapidly.

Crimping the edge of a petal makes it cup-shaped. This shape is a common one. To produce it work around the petal with a knife-blade (not too sharp), holding the knife in the right hand and pressing back small plaits to be firmly held by the thumb of the left hand.

### THIRD LESSON.

GOFFERING. A smooth stick called a goffering tool is used for this purpose. (Fig. 330.) It is used in shaping the inner petals of various double flowers.

The petal is laid on a cotton cushion and the goffering stick drawn sharply through the center, to produce the curled up petals seen in the center of a double rose and other flowers. (Fig. 331, page 183.)

To goffer the edge of a petal lay it upon the cushion as before and draw the tool firmly down from the apex toward the base, and near the edge. Treat the opposite edge in the same manner. (Fig. 332.) Either or both of these processes may be used as the kind of flower requires.

---

### FOURTH LESSON.

CURLING. Though this process is usually performed after the making up of the complete flower, yet the pupil should practice upon separate petals as preparation for perfect work when the various processes have been mastered. By it that natural curl is given to the outer petals which is noticed particularly in roses.

Curling is accomplished by drawing the petals between the thumb and a knife-blade pressed firmly against it. The petal will curl toward the blade. There is danger of cutting the paper if the knife has too sharp an edge.

---

### FIFTH LESSON.

TINTING AND STAINING. The inner point of a flower-petal is often of a different color from the outer part. It is sometimes of a different shade of the same color. These effects may be produced in a simple and satisfactory way by rubbing the center of the square of paper, before folding or cutting, with a dry powder of the shade or color required. A bit of cotton rolled into a small ball may be used for rubbing and blending the powder. It will require some practice and experience to do this delicately.

Staining is done by dipping in water in which a little powder of the color required has been dissolved, the edge or other part of the petal which it is desired to color. After undergoing this process the petal must be allowed to dry thoroughly before crimping or goffering. Some of the common and necessary colors for a beginner's work are Vermilion, Chrome Yellow, Carmine, Pink, Vandyke Brown, Chrome Green, Ivory Black, and White.

### SIXTH LESSON.

THE POPPY. If the pupil has practiced faithfully upon the processes described in the foregoing lessons it is presumable that he is now prepared to attempt the entire construction of a simple flower. The poppy being a common flower and one whose parts are not difficult to imitate and adjust, is a good example for this purpose.

Cut four squares from paper of proper color; fold and cut them into slips of required shape. The two slips designed for the outer part of the flower should be a little larger than the other two. Tint the center with Ivory Black and Vandyke Brown. Crimp the edges of the petals. Select a stem with culot (or, center with stamens). Thrust the stem through the center of one of the smaller slips. Slip this up to the culot and gum it slightly to place. Repeat this with the other small slip, arranging it so that the petals will cover the spaces of first slip. Slip and gum the outer sets of petals in like manner. A few delicate touches will finish and shape the flower nicely.

Arrange seed pod and leaves properly and wind the stem with gummed tissue paper to hold all firmly in place.

It may require several lessons to make a satisfactory flower. A number of flowers of the same kind should be made before a new variety is undertaken by the learner.

### GENERAL SUGGESTIONS.

In briefly describing these chief operations, a general idea of the work of paper flower making has been presented. Practice and experience under wise direction must do the rest. Repeated trial even in the face of continued failure and discouragement must result finally in success. The worker will sometimes fail by being too painstaking. Confidence to work boldly and rapidly will do much to promote easy workmanship and the flower thus made will be more graceful and more natural. We can do no better in striving to imitate Nature than to endeavor to acquire her apparent carelessness.

The instructor will find it an advantage to furnish as models, samples of the various kinds of paper flowers to be made by the pupils. Sacrificing one of these occasionally, by allowing a pupil to take it apart in order to study its construction, will perhaps be found of some assistance. Patterns for petals of different kinds of flowers can be readily drawn from a picture, or from a pattern flower, or better

PAPER FLOWER MAKING.

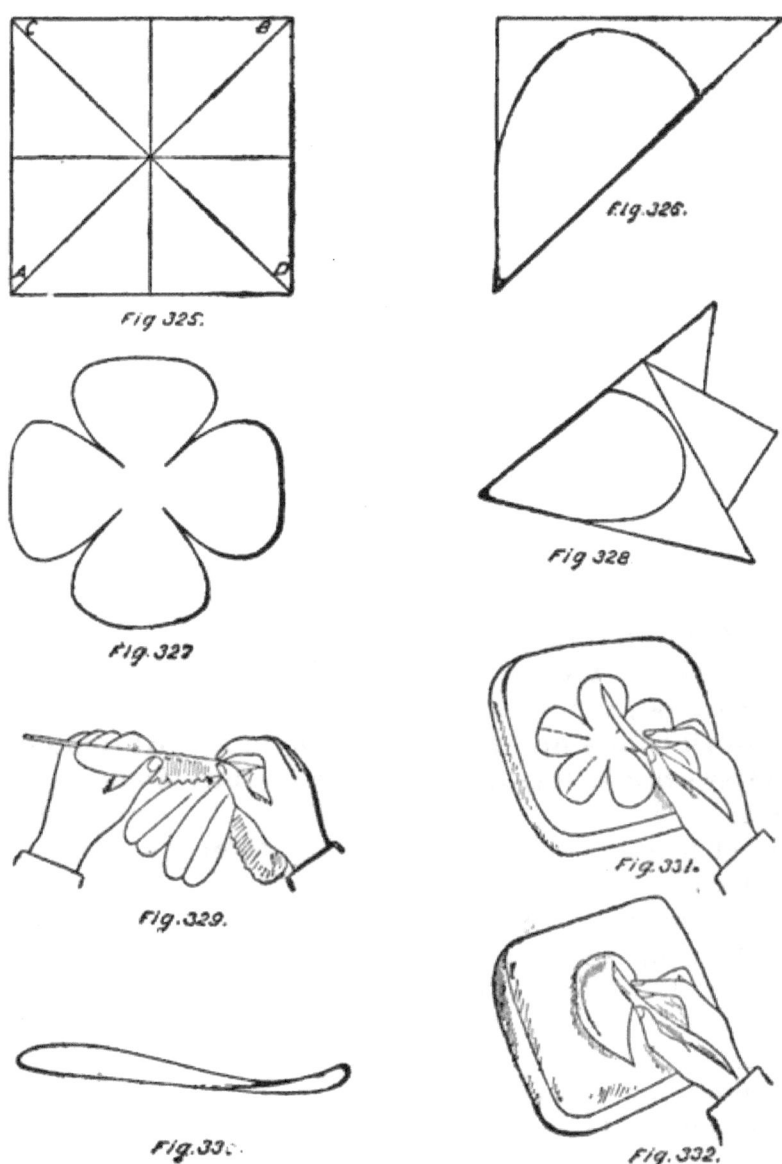

Fig. 325.

Fig. 326.

Fig. 327.

Fig. 328.

Fig. 329.

Fig. 330.

Fig. 331.

Fig. 332.

than either, from the natural flower, if it be procurable.

The most exquisite neatness should prevail at the little flower maker's table or desk. All materials should be handled daintily, that the flowers when completed may show perfect finish. Taste in arranging and grouping bunches and sprays of flowers should receive much attention. They may also be mounted in a number of pretty and effective ways.

The rose, dahlia, daisy, and snowball are good examples of flowers to follow the poppy in successive lessons. After that much has been accomplished, the pupil's choice may guide him in the selection of the flower which he wishes to make.

Each separate lesson in flower making should occupy from fifteen to twenty minutes at least. With advanced pupils more time may be allowed if desired, but the work should never be continued until the pupil regards dismissal with relief.

As a rule, all manual lessons should be cut short when the interest is high; then the pupil will return to the work at the next time with a feeling of pleasure.

## *PENMANSHIP.*

The previous plan is still followed, with the introduction of more advanced Tracing and Copy Books.

## *DRAWING.*

The work of all of the previous years is reviewed, and a Primary Drawing Book presented for pencil-work.

## *GYMNASTICS.*

The previous exercises are continued, and additional exercises.

[NOTE.—The remarks made under Writing, Drawing and Gymnastics, in Part III., Chapter I., should be referred to here.

# PART FOURTH.

## Chapter I.

### *MANUAL WORK.*

FOR THE BOYS AND GIRLS OF THE GRAMMAR SCHOOL—SEVENTH, EIGHTH, AND NINTH GRADES.

PRIMARY METHODS. In learning the Manual Arts in the Primary and Second Grades, boys and girls engage alike, without distinction of sex; a time is provided for the Industrial Class, upon the general program, and it is summoned, conducted, and dismissed in the same manner as that employed in class recitation in the ordinary school curriculum. When accessible, a recitation room may be used exclusively for this work, in order that greater freedom from restraint than that allowed in usual recitations may be accorded to the pupils. But although a separate room is desirable, it is by no means an absolute necessity, since with proper superintendence, Manual Work in the lower grades may be carried on in the general school-room.

A WORK ROOM DESIRABLE. When, however, the pupils have reached the age for promotion to the Grammar School, separate employment for the sexes being furnished, the former plan of operation must of necessity be abandoned, and such a one adopted as will best meet the requirements of the new situation.

To continue the work in the general school-room is obviously impracticable in the case of the boys in their work with tools, and unless the capacity of the shop provided be very large, the boys cannot all find employment there at the same time. But, of course, in case of this being possible, the school room might be abandoned at stated times for the use of the girls, by whom it could be transformed into a temporary sewing room. With the above mentioned provision for the boys, a plan of this nature might be effectively employed, even in a Grammar School of considerable size, the instructor passing from room to room and supervising the work of each in turn.

Another plan, which in the majority of cases will undoubtedly prove the more feasible of the two, is to set apart and fit up two rooms for Manual Work, the one with special reference to the requirements of the girls' occupations, and the other as a workshop in which the boys may learn the use of tools and implements. Each of these rooms should be in charge of an instructor, who may receive classes of convenient size at stated times from the several school-rooms. As one class is dismissed another is received upon the ordinary plan of recitation room work.

In the sewing-room a class of twelve, or at the most fifteen members, can be accommodated and attended by a single instructor. The classes should be arranged, as far as possible, so that each one shall contain pupils engaged in the same occupation.*

THE BEGINNING MAY BE SMALL. The facilities which are at hand or may be afforded for fitting up these rooms may be meager at first, but since the work itself is creative, growth and accumulation are certain to be the fruits of activity and judicious expenditure of moderate or even small capital. There should be a place for every kind of article and material, and the instructor should exercise unceasing care for every appurtenance and watchful supervision over all work and all workers. No work, material, or implement should ever be taken from the room except by permission of the instructor in charge.

Boxes, stands, tables, cupboards, shelves, and many other articles of necessity and convenience will be supplied for the sewing room from the boys' workshop, in a short time after the work has been established.

THE FACILITIES. The sewing room should be properly lighted, heated, and ventilated, and the pupils should be taught to maintain proper position of the body while at work. All rules for the regulation of a well-ordered room used for ordinary school purposes should be carefully observed.

---

*The pupils and the hours are chosen by the class teachers, with direct reference to the best interests of the persons selected, and the class as a whole. In the same way they are selected for the Printing Office. When not members of the Manual Training Class in the school, pupils are encouraged to do as much work as they may be able at home, for which due credit is given them in the course.

It may be well to add here, that the shop and sewing room, both in the event that competent instructors cannot be secured from the outside, may be placed in charge of the older pupils, under the general direction of a regular teacher, who is capable and willing. Or if it becomes necessary, they may be employed to assist the instructor in charge. In either case the expense would be lighter, and generally the work satisfactory.

# Chapter II.

## SUGGESTIONS, LESSONS, AND METHODS OF INSTRUCTION IN MANUAL TRAINING.

THE GRAMMAR SCHOOL—SEVENTH, EIGHTH, AND NINTH GRADES.

1. THE PUPILS. These vary in age, according to the grade they occupy, from nine or ten to fourteen or fifteen years.

2. LENGTH OF LESSONS AND AMOUNT OF WORK. Instruction in the Manual Arts may be given in the Grammar Schools upon every day in the week, or alternate days. Each lesson should be at least thirty minutes in length, but no lesson can profitably occupy a longer time than forty-five minutes. The proficiency acquired by each pupil should determine the number of lessons in any occupation, required in each individual case.

3. STUDIES AND OCCUPATION. A plan of the work deemed requisite for this Department is laid down in the Appendix, Chapter II.

4. THE MANUAL ARTS. A general scheme for work in the various Manual Employments is suggested as follows:—

*For Boys.*

1. Lessons in the use of the Hammer.
2. Drawing Lines and Laying Off Distances.
3. The Saw and its uses.

4. The Plane and its uses.

5. Lessons in the use of the Hammer, Saw, Plane, and Marking Tools.

6. The Bit for Boring.

7. The Chisel and Mortise and Tenon Joint.

8. The Miter Joint, the Dowel Joint, the Dovetail Joint.

9. Printing, Type-setting.

10. Penmanship.

11. Drawing.

12. Gymnastics.

13. Reviews.

### *For Girls.*

1. Sewing Over-and-Over, Running, Hemming, Stitching, Overcasting, Gathering.

2. Knitting.

3. Crocheting.

4. Patching, Darning, and Making Button Holes.

5. Printing, Type-setting.

6. Penmanship.

7. Drawing.

8. Gymnastics.

9. Reviews.

## Chapter III.

*LESSONS IN PLAIN SEWING, RUNNING, GATHERING, STITCHING AND OVER-CASTING.*

#### GIRLS' DEPARTMENT.

PLAIN SEWING. The greater number of the girls entering the Grammar School from the Sixth Grade, will have obtained some previous knowledge of rudimentary sewing, and some may have reached a considerable degree of proficiency. But on account of those to whom further practice is necessary, as well as those who are making their first entrance into the school at the Grammar Department, a review of hemming and over-and-over sewing constitutes the first lessons.

The materials for these lessons, and the manner of conducting the work, are in all essential particulars the same as described in the Sixth Grade. Small bags of unbleached muslin are used, and upon these the girls practice hemming and overhand sewing, taking out imperfect work repeatedly and doing it over and over until the instructor in charge shall pronounce it satisfactory. As soon as this verdict is passed upon the work of any pupil she is permitted to advance to the next class, and begin a new occupation.

Six or eight lessons of forty minutes' duration will usually suffice for this work, and if the previous attainments of any pupil have been good, fewer lessons, in her case, will be required. One or more lessons should be given before the work is relinquished for "over-and-over" sewing on two strips of fine selvage. As a final specimen of work before promotion, a well-constructed piece of patchwork may be presented. For fine hemming and for practice in turning narrow hems, a linen napkin or a breadth of fine ruffling furnishes good material. Each piece of unfinished work should be neatly labeled at the close of every lesson, and all materials deposited in suitable places provided.

A pleasant atmosphere of sociability should pervade the little "Sewing Circle." When not engaged in giving necessary instructions, the teacher should endeaver to promote good culture on the part of the pupils by introducing interesting topics for discussion, or by reading aloud for their instruction or amusement. Unless the class is necessarily conducted in a room where other pupils are engaged in study, communication in a proper measure should not be restricted between the pupils. The pleasure derived from such easy unrestraint will be so highly appreciated, that proper bounds will not often be disregarded if restriction from all conversation be the offender's penalty.

RUNNING. For running may be used at first strips of unbleached muslin, afterward bleached cloth of fine texture, and calico, finally fine cambric. For each kind of fabric suit size of needle and thread to texture of cloth. The pieces first used should be eight inches long by five wide. Patchwork of simple pattern furnishes good work for practice in running. The instructor should make frequent examinations of all work, and advance pupils to new employment as soon as they attain a proper degree of proficiency. Four or five lessons of forty minutes each will suffice for ordinary pupils.

GATHERING. Any of the materials before used in sewing may be employed in teaching gathering. The thread may be used double or single. Laying gathers should be taught as an accompaniment to every lesson in gathering. This work should occupy at least five lessons. Whipping gathers should occupy five more. Then each pupil should make a small article employing all of the kinds of stitches learned. A simple garment for a doll affords good opportunity for this exercise.

STITCHING. The sewing materials before mentioned are again called into requisition for learning stitching. This work is of a more difficult nature than any sewing before undertaken, and the opportunity afforded for the attainment of precise and beautiful workmanship should not be neglected. A fine linen shirt bosom may constitute the material for a final piece of work to be submitted to the examining instructor.

Six or more lessons of forty minutes should be occupied with this work.

OVERCASTING. For learning overcasting the seams previously made in running may be used. The stitches should be taken half the depth of the seam, and as far apart as each stitch is deep. Two or three lessons will be sufficient for this work.

## LESSONS IN CROCHETING.

In the Grammar School the first work in crocheting consists of a complete review of all of the easy stitches learned in the Sixth Grade. Cotton or linen thread may be used, also silk, worsted, and yarn. When the simple stitches are learned more complicated varieties are presented. The Afghan stitch, the Star stitch, and other fancy stitches, and combinations and variations of all the stitches learned, afford occupation for any number of lessons. Several pieces of good crochet work should be required of each pupil before she is permitted to relinquish the work. The numberless beautiful as well as useful articles which can be created, give by reason of their pleasing variety a wide scope for very delightful occupation. Mittens, shawls, hoods, edgings of various kinds, tidies, and infants' sacques, bonnets, shirts, and tiny shoes; these and a hundred other articles of common use make up the list from which the well advanced pupils in crocheting may choose. All work should be performed by the direction and under the personal supervision of the instructor in charge.

Crocheting is pre-eminently an occupation which can admit of no medium workmanship. A slovenly or unworkmanlike piece of crocheting should never be allowed to supply the uses for which the delicately beautiful creations of this graceful art are designed.

For this reason, thorough proficiency should be required in simple crocheting before any attempt at elaborate work is permitted.

## LESSONS IN KNITTING.

All of the simple work before described and illustrated should be reviewed before farther advance is attempted. Casting on stitches, knitting plain and purl, narrowing, widening, and binding off, and knitting very simple edgings on two needles, should constitute the practice-work for many lessons. Then, under the direction of the instructor, the pupils undertake more elaborate patterns of lace, and also learn to knit a stocking and a mitten, using four needles. Many of the articles mentioned under the head of Crocheting can be produced by knitting.

The number of lessons required by the pupils in both Crocheting and Knitting, depends very much upon the natural adaptability which they may possess for the work. Thorough proficiency in the work in hand should constitute in each individual case the only condition upon which promotion to an advanced employment is dependent.

---

## LESSONS IN MENDING, PATCHING AND DARNING, MAKING BUTTON HOLES.

MENDING. To learn to mend neatly a torn or worn garment is certainly an acquisition which every girl should desire to possess, and although as little pleasure is anticipated from it, perhaps, as from any one of the Manual occupations, there is none that affords greater satisfaction from accomplished results.

A finished piece of handsome embroidery is not contemplated with more pride and gratification by the youthful needlewoman than is a perfectly set patch or difficult piece of darning produced by her newly skilled fingers.

Perhaps the proverbial "stitch in time" would not be so often neglected if the reclaiming power of the darning needle were more generally appreciated. Intelligent pupils will not find it difficult, after a little experience, to understand that the very commonplace occupation of patching or darning may be regarded from an artistic stand-point. The disfavor with which "mending day" is accustomed to be regarded in homes where a wise maternal government decrees that each daughter shall perform her own mending service, might be greatly lessened if the idea of elevating a humble task by the mastery of perfect workmanship were cultivated and made to dominate over the feeling which is born of doing a distasteful thing because it is necessary.

PATCHING AND DARNING. The materials employed for this work consist of pieces of woolen and cotton goods at first, and afterward silk, linen, and finer fabrics. Needle and thread should suit the material.

The pupils should learn to trim the rent neatly; to cut and fit a patch and baste it in; to stitch it; and finally to press the finished work. A patch may also be sewed over-and-over and hemmed in. The instructor should impart to the pupils all of the well known directions necessary for observance in skillful patching.

For darning the same materials are used as for patching, except in two or three final lessons, which should be devoted to stocking darning.

Both of these occupations afford most excellent opportunity for the display of nice workmanship. Five lessons should be devoted to patching, five to plain darning and five to stocking darning.

MAKING BUTTON HOLES. For button holes, use the same materials as for patching and darning. The first practice should be in marking, cutting, barring and overcasting, by direction of the teacher in charge. At least twelve good button holes should be required on cotton fabric and twelve on woolen before the work may be considered completed. The number of lessons necessary will depend upon the aptitude of the pupil. The examining teacher should estimate the quality of a button hole by a standard little short of perfection.

## Chapter IV.

## *THE SHOP FOR WOOD-WORK.*

BOYS' DEPARTMENT—INTRODUCTORY.

THEORY AND PRACTICE. A practical knowledge of the use of wood-working tools is educational. A theoretical knowledge of their use is better than none at all. A simple study of the construction of the various implements employed in working in wood, the purposes to which they may be devoted, the ends that may be accomplished by an intelligent use of them, and the part they play in producing the conveniences, the luxuries, and the progress of the civilized world, is well worth much valuable time, and the best efforts of all the youth in the land, and of many mature minds as well. Elementary theoretical mechanics might well be substituted for subjects of less importance in our secondary schools. They would do much toward awakening an interest in school work, and render useful aid in gaining a practical knowledge of other subjects of study. They would furnish the mind of the pupil with convincing evidence of the uses of all knowledge. It would give them new ideas which would serve as the basis for higher aspirations and nobler lives.

When, however, the youth enters the shop and learns by actual practice how to use these implements, and makes and builds with them, becoming thereby a creator of values, there is scarce a limit to the benefits that may be realized if the work be judiciously and persistently carried on. This kind of manual training not only develops the muscle, but wakes up the mind, creates a lively interest in all the methods of obtaining knowledge, gives added zest to many experiences of life and promotes the self-respect of the youth, without which there can be no real progress.

It will be the purpose of the lessons which follow, to give the names of the tools and implements used in wood-work, the parts of which they are composed, the various uses to which they may be put, practical lessons in using them, and exercises in making and constructing articles of use and beauty. After the pupil has learned how to use a few of the tools, he will have the privilege of making objects to test the accuracy of his knowledge and the facility he has acquired in the use of the tools thus far placed in his hands. These exercises will be given at intervals all through the course of instruction.

THE SHOP. The shop may be any room which can be (1) well-lighted, that is, the light so diffused as to be soft and at the same time so clear that the worker can distinguish accurately the minute features of his work. (2) The temperature should be so regulated that the occupant will be in no danger from change, whether exercising violently or moderately. (3) Ventilation must also be looked after. As the exercise of the pupil therein is to be invigorating, strengthening, health-giving, the air taken into the lungs must be fresh and pure.

Light, heat, and ventilation should be watched with as much interest for the welfare of the pupil in the shop as in the school-room.

There should be room in the shop for workbenches, lathes and tools, as many as are to be provided and used, and they should be so placed as to insure the best position for the convenience of the workers. A little thought on the part of the foreman or instructor, whoever he may be, given to the features named above will be of great service in making the shop just what it ought to be, a pleasant, attractive resort of the young learner.

THE WORK-BENCH. The work-bench may be stationary, portable, or knock-down according to circumstances. The first is placed against the wall of the room, and is sometimes fastened to it, on which also many of the tools are suspended. The second does not differ materially from the first, except that it stands clear from the wall, and if wide enough may serve as a double bench. The knockdown is made by placing a plank fifteen to twenty inches wide and six to eight feet long upon two wooden horses and nailing strips of board on the under side of the plank to hold it in place. The bench-vise must be attached to the forward horse. This bench can be taken down in a moment, and laid aside, to be set up again at will, in the school room for after school work if desirable. The portable bench may have a deep drawer in which to keep the tools, instead of a tool-chest, or a shelf under the bench will answer the same purpose and save room.

TOOLS FOR WORKING IN WOOD AND HOW TO USE THEM. The object of the exercises and lessons hereafter given is to show the learner by actual practice how to use the various implements commonly employed in working in wood. The first lessons will be explained somewhat minutely in order that the young workman may clearly understand what he is to do, but he will be left more and more to rely upon his own judgment as to the meaning of the terms used. Each tool will be taken up by itself, beginning with those that are in common use and easily understood.

THE HAMMER. The hammer is a striking tool. There are several of the same general character. Among these may be named, the tack-hammer, the nail-hammer, the mallet, the beetle, the hatchet, the adze, and the axe. In this chapter instruction will be given in the use of the nail-hammer only. Three kinds of blows may be struck with it,—the light, the medium, and the heavy blow.

The hammer has a handle and a head. The head has an eye into which the handle is fastened, the face, which meets the object struck, and the claw for withdrawing the nail. Hammers are numbered according to size, No. 1 being the largest. No. 2 or 3 is large enough for boys in the Grammar School. A nail-set is frequently used with the hammer, therefore one should be supplied with each hammer.

MATERIAL. Each member of the class will require three or four pieces of pine, six feet in length, six inches wide, and one inch, or seven-eighths of an inch thick, and some pieces of hemlock to serve as joists in blind-nailing, also one-half pound each Nos. 3, 4, and 6 nails.

### FIRST LESSON.

LEARN TO STRIKE THE LIGHT BLOW. Stand with the right side to the bench, on which is placed a piece of soft board. Take the handle of the tool firmly in the right hand. Close the fingers over it and extend the thumb along the side. Let the face of the tool rest on the board, and the forearm just above and parallel with the upper surface of the bench. (Fig. 334, p. 205). Now raise the hammer with the wrist movement, keeping the forearm in the same position, and strike light blows in this way until the lesson is fully mastered. (Fig. 335.)

### SECOND LESSON.

LEARN TO STRIKE THE MEDIUM BLOW. Stand beside the bench, holding the hammer as in Lesson 2. Now raise the hammer with the wrist as before, and also the forearm. (Fig. 336.) Keep the arm above the elbow from moving, and strike until the full medium blow can be given with ease and precision.

### THIRD LESSON.

LEARN TO STRIKE THE HEAVY BLOW. Standing as before lift the hammer, using the wrist, and the elbow, and also the shoulder. (Fig. 337.) Move the arm at the shoulder, thus raising the hammer high, and strike until the blows can be struck with ease, precision, and great force.

### FOURTH LESSON.

LEARN TO STRIKE SQUARE BLOWS WITH PRECISION. Lay a piece of pine board (or other soft wood) on the bench. Strike light blows about one inch apart in lines over the surface. In the same places strike medium blows, and then heavy blows, striving to sink the face of the hammer evenly into the surface of the board. (Fig. 338.)

---

### FIFTH LESSON.

LEARN TO DRIVE A NAIL BY FIRST PROCESS, OR STRAIGHT NAILING. Lay a piece of board of pine or other soft wood on the bench. With a square and pencil draw a line and on it make points one-half inch apart. Take a number 4 or shingle nail between the thumb and first finger of the left hand, hold it upright with the greater width of the point across the grain of the board. (Fig. 339.) Strike one or two light blows; then remove the hand and drive it home. In the same way, drive a nail at each point in the line.

---

### SIXTH LESSON.

LEARN TO NAIL WITH A NAIL-SET. Drive a row of nails as before, taking care not to mark the board with the face of the hammer. Then with a nail-set, sink the head of the nails below the surface of the board.

---

### SEVENTH LESSON.

LEARN TO DRIVE NAILS IN A LINE. Lay off a line and divide it into spaces one-half inch apart and at each point drive nails that shall be in exact line with each other.

#### EIGHTH LESSON.

LEARN TO DRIVE NAILS FLUSH. Lay off a line as before and at each point drive a nail, leaving the head a little above the surface of the board. See that all are the same height and in a straight line.

---

#### NINTH LESSON.

LEARN HOW TO DO BLIND-NAILING. This kind of nailing is used generally in laying floors. Lay two strips of board on the bench that shall serve as joists. Across these place a piece of board for the flooring. To keep it from moving, drive a nail or two against the back side of it. Then on the other side, a little below the upper edge, drive a nail obliquely into the board and through into the joist. Against this piece now place another and nail it to the joist as before. (Fig. 340.)

---

#### TENTH LESSON.

LEARN TOE-NAILING. Lay a piece of board on the bench, and then place on this board another on its edge. Drive a nail obliquely into this board and through into the board on which it rests. (Fig. 341.) Nail both sides, and see that the upper board is held firmly in its place.

---

#### ELEVENTH LESSON.

LEARN TO DRIVE NAILS HORIZONTALLY AT THE RIGHT, AND AT THE LEFT HAND. Fasten a piece of board in the bench-vise, extending it along against the side of the bench. Into this board drive nails

## THE HAMMER. 205

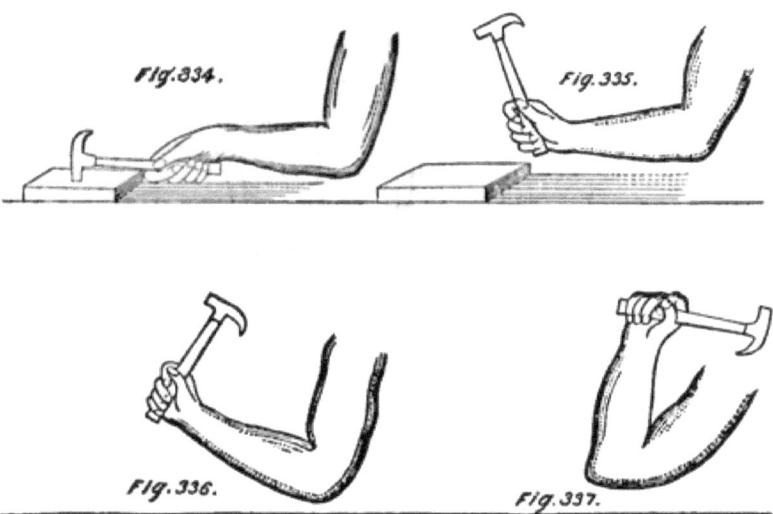

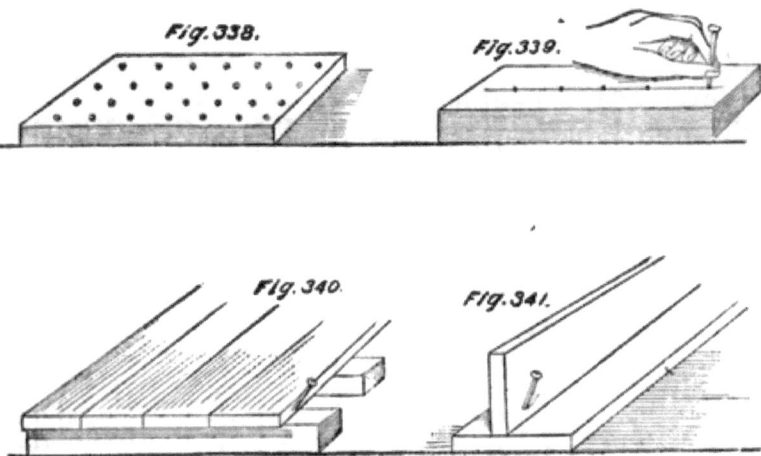

horizontally, standing first with the right side to the bench, and then the left side. (Figs. 342, 343, p. 214.) The seeming awkwardness of the position will be overcome by successfully driving a few nails.

### TWELFTH LESSON.

LEARN TO WITHDRAW NAILS. In order to use the claw in withdrawing a nail, the nail must first be started, so as to raise the head a little above the face of the board; then apply the claw. If the head of the nail is some distance above the board, place a small block under the hammer, and then the nail can be withdrawn without bending it. (Fig. 344.)

## *DRAWING LINES AND LAYING OFF DISTANCES.*

THE TOOLS. The rule, the square, the try-square, the ten-foot pole (it may be less), the gauge, the awl, the knife and pencil, and the chalk and line. These instruments are all used more or less in drawing lines and laying off distances. It is not necessary to describe them at this time. Brief object lessons should however be given by the foreman or instructor in order that their construction and parts may be thoroughly understood by the young workman.

MATERIAL. For each member of the class there should be two boards of pine, or other soft wood, ten feet or twelve feet long, ten inches wide and one inch or seven-eighths of an inch thick. Get planed lumber (at least, one side,) if convenient, and with one straight edge. The lumber for these lessons can be further used for sawing and other purposes.

### FIRST LESSON.

LEARN TO DRAW A STRAIGHT LINE ON THE SURFACE OF A BOARD. Lay the board on the bench; with the pencil mark two points at greater or less distance from each other; close beside these two points lay the straight edge of the rule, square, or pole; upon it place the thumb and fingers in opposite directions so as to hold it firmly in position (Fig. 345, p. 214), then connect the two points by drawing the pencil along the straight edge between them. To learn to hold the straight edge in position with facility is the object of this exercise.

### SECOND LESSON.

LEARN TO LAY OFF DISTANCES AND CONNECT THEM WITH A STRAIGHT LINE. Place the graduated edge of the rule or other measure upon the board in the direction desired; with the thumb and fingers of the left hand hold it firmly in position; then mark the two points to be connected, with the awl or pencil, and connect them with a line drawn along the graduated edge.

### THIRD LESSON.

LEARN TO CONNECT TWO GIVEN POINTS ON A BOARD BY A CHALK LINE. Through the loop in the end of the line pass the awl and thrust it into one of the points; hold the line tight in the left hand; take the chalk in the right hand; press it against the line, covering it with the thumb, drawing it forth and back until the distance to be laid off is chalked. Now with the line still drawn tight, place it upon the surface of the board connecting the points, and with the right hand raise the line vertically and let it go. On removing the line the chalk mark will be seen. The straightness of the chalk line will depend upon holding the line tight and raising exactly in a vertical direction. The chalk line is generally used for distances too great to be covered by the rule or square.

### FOURTH LESSON.

LEARN TO DRAW A LINE ON THE SURFACE OF A BOARD PARALLEL TO THE EDGE. Place the marking-tool, which may be the pencil, the knife, or the awl, upon the point at one end of the line. Take the rule in the left hand, enclosing it with the fingers and the thumb on the upper side. Place the rule upon the board, the end touching the marking-tool, its length square with and the forefinger against the edge of the board. Let the second finger of the right hand rest against the end of the rule just behind and touching the pencil.

Now move the marking-tool and the rule along the surface to the end of the line. This method of ruling may be used where great accuracy is not required.

### FIFTH LESSON.

LEARN TO DRAW A LINE PARALLEL TO THE STRAIGHT EDGE OF A BOARD, AT A GIVEN DISTANCE FROM IT, WITH THE GAUGE. Set the gauge at the required distance. Take it in the right hand, the thumb opposite the spur, the forefinger over the head, and the remaining fingers closed over the bar. Place the head of the gauge against the straight-edge, the bar resting on the board, so that the spur will just touch the surface. Move the gauge to the initial point of the line and then push it forward, turning the gauge so that it will mark the line called for.

### SIXTH LESSON.

LEARN TO DRAW A LINE THROUGH A GIVEN POINT, AT RIGHT ANGLES TO A STRAIGHT EDGE OF A BLOCK OR PIECE OF BOARD. Place the try-square upon the board, the beam pressed against the straight edge, and the blade on the surface. Put the marking-tool upon the given point, and move the try-square so that the blade will touch it, but not cover it. Then draw the marking-tool along the edge of the blade across the board, holding it in the same direction the entire distance. If the straight edge is on the opposite side of the board, put the gauge on that side and adjust it, and then draw the line as before.

### SEVENTH LESSON.

LEARN TO DRAW A CURVE ON THE SURFACE OF A BOARD. (The curve will be part of the circumference of a circle.) At a point chosen for the center, fasten the end of a line by means of a loop and an awl. At a given distance from the center grasp the line and marking-tool together, holding them firmly with the point of the tool on the surface of the board. Then with the line drawn tightly move the point on the board. It will describe the curve required.

## SAWING.

THE TOOLS. The name of the implement used in sawing is the saw. There are a variety of tools of this kind. They are divided first into circular and straight saws. The former are called circular, band, and scroll saws. They are generally run by steam, water, horse, or foot power. The latter are named the cross-cut and the hand saws. The cross-cut is a large saw with a handle at each end, and run by two men at the same time. The hand saws are named the cutting-off saw, the rip saw, and the back saw. A pair of saw-horses will be required for the following lessons, to accompany each work-bench, also a bench hook and a miter box. The object of this chapter will be to illustrate the use of the hand saws.

MATERIAL. The lumber used in the lessons for drawing lines and laying off distances will be sufficient for the following lessons. The lines already drawn may be followed or new ones made.

---

#### FIRST LESSON.

LEARN TO CUT OFF THE END OF A PIECE OF BOARD AT RIGHT ANGLES TO THE STRAIGHT EDGE. Lay the board upon the two horses, and draw a line through the point at which the board is to be cut off. (See Page 209, Lesson 6.) Take the cut-off saw in the right hand, extending the forefinger along the handle and closing the other fingers through the opening. This grasp will help to hold the saw steady in the hand. At the line on the straight edge, place the saw at about the middle of the cutting edge. Take hold of the straight edge of the board with the left hand, resting the thumb nail gently against the blade of the saw. Now draw the saw lightly toward the shoulder, then reverse the motion and continue to move the saw back and forth.

Care must be taken not to tear the edge in starting the saw. Keep the saw close to the line but not entirely effacing it. To determine whether the saw is running at right angles with the face of the board, place the back of the beam of the square on the face of the board, the arm extending upward. Shove the square against the blade of the saw and if it is cutting square with one face the arm will touch the saw its full length.

When the saw has cut nearly through, place the left knee upon the board to hold it steady and with the left hand grasp the edge both sides of the cut or kerf. The strokes of the saw should now be short, light, and quick to the finish. In this way the piece will be completely severed without slivering or splintering the edge.

### SECOND LESSON.

LEARN TO CUT A STRIP FROM A BOARD LENGTHWISE, ON A LINE DRAWN FOR THAT PURPOSE. Lay the board on the horses and draw the line, using the square, the pole, or the chalk and line, as the length may require. Place the knee upon the board to keep it in position. Take the rip saw in the right hand, and the end of the board in the left; then start the saw as directed in Lesson I. Follow the line closely but not entirely effacing it.

If the saw should bind in the kerf, as it sometimes does in sawing on a long line, insert a wooden wedge or a scratch-awl to keep the sides apart.

After sawing to near the first horse draw the board backward so as to bring the saw between the horses; and if the cut is to be very long, it may be necessary to move the board and finish behind the second horse.

### THIRD LESSON.

LEARN TO CUT OFF A PIECE OF BOARD OR BLOCK OF WOOD WITH THE BACK SAW AND BENCH HOOK. The bench hook consists of a strip of board eight to twelve inches in length, with a square block attached to the under side of one end and the upper side at the other.

Place it upon the bench, with the block on the lower side against the front side of the bench extending toward the opposite side. Against the hook at the other put the piece of board or block of wood with the line at which it is to be cut off just at the right. With the left hand hold the wood firmly in position, and start the saw nearly as directed in Lesson I.

As the saw will rest on the wood it must be started with care, running it very lightly and cutting the kerf vertically.

### FOURTH LESSON.

LEARN TO CUT OFF A PIECE OF BOARD OR BLOCK OF WOOD WITH THE BACK SAW AND THE MITER BOX. The miter box is made of a board one inch in thickness and one or two feet long with side pieces rising to the width of the saw used with it. Square across the box kerfs are cut to the surface of the bottom. They may also be cut at any angle across the box. In the box place the board or block to be cut off with the line at which it is to be cut exactly in line with the saw cuts. Hold the board or block firm in its place with the left hand, and with the right place the saw in the cuts, drawing it back and forth lightly until the object is cut through. At this point the saw will cease cutting and run resting on the back, upon the sides of the box.

#### FIFTH LESSON.

LEARN TO CUT OFF A PIECE OF BOARD OR BLOCK OF WOOD AT ANY ANGLE WITH THE BACK-SAW AND MITER BOX. The saw cuts must first be made in the miter box, at the angle desired. The work of cutting off the object is conducted the same as in Lesson IV., using the saw cuts making the angle indicated.

### *PLANING.*

THE TOOLS. These are the plane, which is called by different names, the principal ones being the jack-plane, the smoothing-plane, and the jointer. The jack-plane is used to reduce the rough surface of a board. It is usually fifteen to eighteen inches in length. It is followed by the smoothing-plane, which still further reduces the surface. It leaves the surface quite even and smooth and ready to be used or worked up. It is eight to ten inches long. When the edges of two boards or other material are to be put together, or when put to other uses it is often

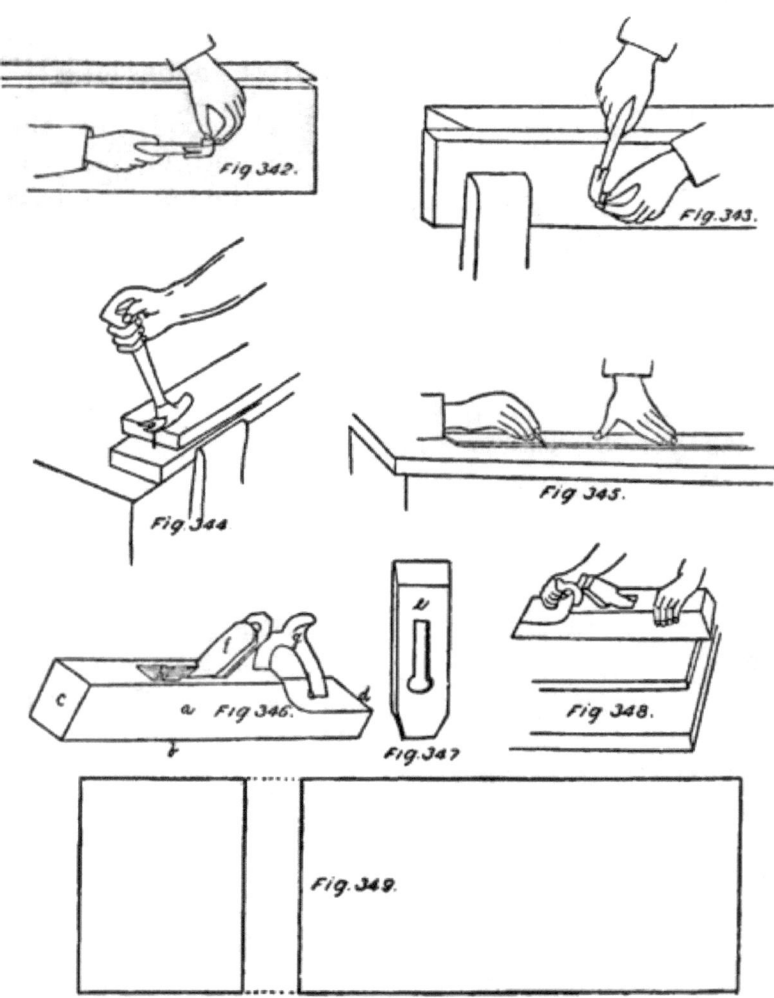

necessary that the edges should be very straight. To accomplish this work is the office of the jointer. We have the short, the medium, and the long jointers. They are twenty to twenty-five inches long.

THE PARTS OF THE PLANE. The body of the plane is called the stock ($a$); the bottom, the sole ($b$); the front end, the toe ($c$); the hind end, the heel ($d$); the cutting blade, the plane-iron ($e$)—(which sometimes has an iron cap, fastened to it with a screw); the chip of wood, which fastens the iron in the plane ($f$); and the handle ($g$). The opening in the stock for receiving the iron is called the throat ($h$). (Figs. 346–347, p. 214). An oil-stone will be required for the lessons in this chapter.

MATERIAL. Supply each member of the class with two pine (or other soft wood) boards, six feet long, six inches to eight inches wide and one inch to seven-eighths of an inch thick. For the purposes of planing, the lumber should be free from knots, and otherwise of good quality. It will be well also to give the pupil a piece of hard wood (ash, oak or chestnut) to reduce with the plane. Pieces left from other lessons can generally be utilized to advantage.

---

#### FIRST LESSON.

LEARN TO REMOVE THE PLANE-IRON FROM THE STOCK. Take the plane in the left hand with the fingers clasping the under side and the thumb in the throat pressing lightly against the chip. Strike a few light blows upon the toe of the stock, to loosen the chip and the iron. Remove them from the throat, and place them on the bench, together with the stock.

## SECOND LESSON.

**LEARN TO SHARPEN THE PLANE-IRON.** Place an oil-stone upon the bench in convenient position. Remove the cap from the plane-iron if it has one. Take the plane-iron in the right hand, with the thumb on one edge, the third and fourth fingers closed over the other edge, and the first and second fingers extended on the upper side. Place the angular surface of the iron on the oil-stone and rub it over the surface. To assist in moving the iron over the stone, put the fingers of the left hand on the upper surface and the thumb against the opposite side. This will steady the movement and strengthen the pressure. Great care must be taken to keep the angle of the iron with the stone constantly the same, which is about thirty-five degrees, otherwise the angular surface will be oval or irregular, and the bit of the iron will not be sharp. If the bit becomes wire-edged, turn the iron over flat upon the stone, and rub it back and forth a few times. This will remove the wire edge.

## THIRD LESSON.

**LEARN TO REPLACE THE IRON IN THE STOCK.** Take the stock in the left hand with the fingers clasped under the sole, the second finger being over the mouth and the thumb in the throat. Fasten the cap to the iron if it has one. Insert the iron so that the bit or cutting edge will barely touch the second finger over the mouth. Over the iron replace the chip and then drive it so as to hold the iron firmly in its place.

### FOURTH LESSON.

LEARN TO ADJUST THE PLANE-IRON SO THAT IT WILL CUT A SHAVING OF THE SAME THICKNESS THE ENTIRE WIDTH. With the iron replaced take the plane in the left hand with the toe toward the eye so as to sight lengthwise over the sole. If the iron projects evenly beyond the sole, the shaving will be of the same thickness the entire width. To make it project evenly, strike a light blow with the hammer on the side with the greater projection. Continue sighting and striking light blows until the bit or cutting edge of the iron projects evenly through the sole. Then tighten the chip.

### FIFTH LESSON.

LEARN TO ADJUST THE PLANE-IRON FOR CUTTING A THICK OR THIN SHAVING. Take the plane in the hand as in Lesson IV., and sight lengthwise of the plane. If the bit does not project far enough, strike light blows with the hammer on the end of the iron until it does. If it projects too much, strike light blows upon the top of the toe until the bit is withdrawn to the right position. After each adjustment tighten the chip. Also test the adjustment by trying the plane on a board, and readjust until the desired thickness of shaving is received.

### SIXTH LESSON.

LEARN TO REMOVE THE ROUGHNESS FROM THE SURFACE OF A BOARD WITH THE JACK-PLANE. Lay the board on the bench, pressing the forward end sharply against the bench-dog to hold it in position.

Take the handle of the plane firmly in the right hand. (The forefinger is usually extended so as to touch the iron.) Stand with the right side to the bench, at the rear end of the board, and the left foot advanced twelve to eighteen inches. Turn the shoulders toward the bench, and take the forward part of the stock in the left hand with palm down, the thumb on the left side and the fingers on the right side. Place the sole of the plane on the board with the iron just back of the end. Press down with the left hand so as to keep the sole of the plane flat on the surface of the board. Now push steadily and strongly with the right hand. Near the end of the stroke drop the left hand to the side and finish it with the right hand. As the stroke is finished, tip the plane upon the left hand edge of the sole and return it to near the place of beginning. This will prevent dulling the iron. Repeat the strokes until the entire surface has been gone over, advancing the length of a stroke at a time, if the board is too long for one stroke.

### SEVENTH LESSON.

LEARN TO KEEP THE BOARD OF THE SAME THICKNESS THROUGHOUT THE ENTIRE SURFACE WHILE PLANING IT. This must be done by keeping the sole of the plane flat upon the surface of the board. When planing near either edge, be careful not to allow the plane to tip over the side, nor permit it to run too much off the surface, as it will then be more liable to tip. When the plane is at the rear end of the board press heavily with the left hand, and when near the forward end, press heavily with the right hand, to keep it from tipping endwise.

### EIGHTH LESSON.

Learn to Test the Evenness or Flatness of a Plane Surface. Place the plane on the board so that one edge of the sole will rest on the surface and look between the surface and the edge. If it touch the entire length of the plane the surface is even. Move the plane about, over the surface, also sight in different directions, over the surface. A rule with a straight edge or a square may be used instead of the plane. (Fig. 348, p. 214.)

### NINTH LESSON.

Learn to Use the Smoothing-plane to Improve the Surface as left by the Jack-plane. Grasp the smoothing-plane (which has no handle) by the heel with the right hand, the thumb on the left side and the fingers on the right. The palm will then press against the upper part of the heel. Clasp the left hand over the toe, the thumb and first two fingers on the upper side and the other two over the end. Make shorter strokes than with the jack-plane, holding the plane with both hands, in making the stroke and returning the plane. Go over the entire surface of the board. Test the evenness of the surface as directed in Lesson VIII.

### TENTH LESSON.

Learn to make the Surface of a Board Smooth and Even with the Jointer. The plane-iron should be made very sharp, and set so as to cut a very thin shaving, and of the same thickness the entire width. Grasp the plane in the same way and make the strokes in the same manner as with the jack-plane. Go over the surface carefully, removing all ridges and uneven places. Test the condition of the surface as directed in Lesson VIII.

### ELEVENTH LESSON.

LEARN TO STRAIGHTEN THE EDGE OF A BOARD WITH THE JOINTER. Fasten one end of the board in the bench-vise, the other end resting on a bench-pin. The edge to be straightened should of course be uppermost. Start the stroke at the rear end of the board and continue it through to the other end. Continue the strokes until a shaving is cut the whole length of the board. Make tests with the try-square to see if the edge is at right angles to the surface. This is done by placing the beam of the try-square against a planed surface of the board, the blade resting on the edge. Move the square along the board, and if the blade touches the edge the entire width the object is secured.

## *USE OF THE HAMMER, SAW, PLANE, AND MARKING TOOLS.*

TOOLS. The screw-driver should be added to the tools already used.

MATERIAL. Each member of the class should be supplied with a soft wood plank, eight feet long, six inches wide, and one and one-half inch thick, a soft wood board, ten feet long, ten inches wide, and one inch thick, and another of the same dimensions, but one-half inch thick, also three or four sizes of nails, one-quarter pound of each size, and a few pairs of brass hinges and screws for them. In ordinary material it is not necessary to be exact, as whatever is left over can be utilized in lessons to be given afterward.

### FIRST LESSON.

Cut with the saw from a clear pine board, planed on both sides, one inch or seven-eighths inch thick, two pieces four inches by six inches. Note: As these pieces are to serve as the sides of a box, an allowance of one-eighth inch must be made for planing the edges.

### SECOND LESSON.

Cut in the same manner from the same or a similar piece of board, two pieces, four inches by four inches. Note: As these pieces are to be the ends of the box, the same allowance must be made for planing the edges. When exact working measurements are given, this allowance should always be made.

### THIRD LESSON.

Cut from the same or a similar piece of board, a piece four inches by six inches. Note: As this piece is to serve as the bottom of the box, allowance must be made in cutting it for the thickness of the two sides.

### FOURTH LESSON.

Construct the box by nailing the five pieces together. Note: First start three No. 6 nails at each end of one of the side pieces. Place upright the two end pieces, then rest the side piece on them, fitting one end of it to one of the end pieces, and drive the nails. Fit the pieces and drive the nails at the other end. Turn it over and nail on the other side. Finally nail the bottom in place.

### FIFTH LESSON.

From the same or a similar piece of board, construct a box the inside measurement of which shall be four inches by four inches by eight inches. Note: In cutting the pieces be careful to make all due allowances.

### SIXTH LESSON.

From a board three-eighths or one-half inch in thickness make a box of the same inside measurement as given in Lesson V., with a cover and hinges, using sandpaper to finish.

### SEVENTH LESSON.

From a piece of pine or cucumber (whitewood), one and one-half or two inches thick, make two pen-handy-blocks eight inches long by two inches in width. Bevel the upper corners of the sides and ends. With a one-quarter inch bit bore two rows of seven holes each, one inch deep, the rows one-half inch from the edges of the block. It will then be ready for use.

### EIGHTH LESSON.

Make a working drawing for a book-holder, the base to be sixteen inches long and eight inches wide; the ends to be eight inches wide and six inches high.

Scale one-quarter inch for one inch. (Fig. 349, page 214.) Note: Tack a piece of drawing paper smoothly to a drawing board. The board should be at least eighteen inches square and one-half or three-eighths inch thick, and the paper should nearly cover the whole of the upper surface; at least large enough for the drawing.

With a rule or square draw a line with the pencil, four inches long. At each end draw a perpendicular line two inches long. Connect the upper ends of these lines. Near this figure draw another two inches by one and one-half inch. These figures will represent the parts.

### NINTH LESSON.

From a board of pine or other soft wood one-half inch thick, cut with the saw, the base and the two end pieces for a book-holder, on the scale indicated in Lesson VIII. Plane the edges and smooth the ends with sandpaper; and then put the parts together with brass hinges, set into the wood, and the end pieces resting on the ends of the base.

### TENTH LESSON.

Make a working drawing of a saw-horse, to be two feet six inches long, six inches wide, and about twenty inches high; the legs to be six inches wide. Scale one-quarter inch for one inch. Note: The drawing will consist of two figures, one seven and one-half inches by one and one-half inch for the top board, and one five and one-half inches by one and one-half inch for the legs.

### ELEVENTH LESSON.

From a plank of planed pine or other soft wood one and one-half inch thick, cut with the saw the top board, and from a board one inch thick cut the four legs as indicated in Lesson X.

### TWELFTH LESSON.

With the material as prepared in Lesson XI., construct the saw-horse. Plane the edges and the ends of the top board and the legs. Make the sockets in the top board for the legs four inches from the ends. As the legs should slope out a little, the cut in the lower side must be at least one-eighth of an inch less than on the upper side. Nail the legs firmly in place; bevel the lower ends to make them fit evenly and firmly on the floor; and with a plane or saw make the upper ends even with the top board. To keep the legs from spreading apart fit a piece of board six inches wide across the legs under the top board and nail it firmly in place. Note: In planing across the grain never plane to the end at first, as it would chip the corners and spoil the piece. To avoid this keep reversing the piece, planing in one direction and then in the other.

### THIRTEENTH LESSON.

Make a working drawing of the parts of a small wash-bench. Scale one-quarter inch to one inch. The parts are: a top piece twelve inches by eighteen inches; the two ends, each nine inches by seventeen inches. A support across the back two inches by fifteen and one-half inches; and a drawer four inches by six inches by nine inches.

### FOURTEENTH LESSON.

With the same scale as in Lesson XII., draw a front elevation of the wash-bench, showing the position of the drawer, and also of the support.

### FIFTEENTH LESSON.

Cut with the saw from a board of pine or other soft wood one inch thick, the parts of the wash-bench, using one-half inch material for the drawer.

### SIXTEENTH LESSON.

With the material as prepared in Lesson XV., construct a wash-bench, and finish it with sandpaper. Note: The drawer should be placed under the top board at the right, and rest on projecting edges, one nailed to the end board for the bottom to rest on, and the other to the under side of the top board. They must be fitted so that the drawer will run smoothly and closely.

### OTHER LESSONS.

If the circumstances are favorable, each member of the class may now be permitted to make any article at will (one or more), provided it is suited to his capacity and warranted by his experience with the tools already in use.

## *BORING.*

THE TOOLS. The instruments used in boring are the boring-machine, the common auger, and the brace-bits. The boring-machine is used in boring holes in large timber for building and other purposes. The common auger is used for boring holes one inch in diameter and larger, and generally in framing heavy timbers. For boring holes an inch or less in diameter, the brace-bit is the instrument now universally employed. There are several kinds of bits, viz.: the auger bit, with the auger point, varies in size by four-sixteenth inch, from four-sixteenth inch to sixteen-sixteenth inch; the gimlet-bit, with the gimlet point, used in boring holes less in size than one-quarter inch; and the drill bit, with the drill point, used in boring in hard wood, and usually of sizes less than one-quarter inch. The brace, or bit stock, a curved instrument of iron or wood for holding and turning the bits.

MATERIALS. Supply each member of the class with two or three blocks of soft wood one foot by six inches by one-half inch or two inches, also several pieces of board one foot by ten inches by one inch. These pieces can usually be found among the remnants of the material already used.

#### FIRST LESSON.

Learn to bore perpendicular holes with the gimlet-bits. Place a board of soft wood an inch or less in thickness upon the horses. Fasten one of the larger gimlet-bits in the bit-stock. Take the head of the bit-stock in the left hand, closing it over the head; place the bit-stock upright, the point of the gimlet resting on the board. Lean lightly on the left hand, and with the right hand turn the curve of the brace; the gimlet will penetrate the wood. To withdraw the gimlet reverse the motion of the brace. The direction and straightness of the hole will depend upon the holding of the head of the bit-stock in the left hand. It must be held true and steady. Bore several holes.

### SECOND LESSON.

Learn to bore horizontal holes with the gimlet-bits. Fasten a piece of board an inch or less in thickness in the bench-vise so that it will be above the bench several inches. Take the brace with the bit attached, as before directed; place the point of the gimlet against the board, and pressing lightly turn the curve of the brace. In this way bore several holes, taking care to hold the bit-stock steady and true.

### THIRD LESSON.

Learn to bore holes one inch in diameter, one inch deep in a pine or other soft wood block. Lay the block on the bench or horses; draw a line along the center of one of the surfaces, and mark points on this line at intervals of one and one-eighth inch. Then fasten the block in the bench-vise, or on one of the horses by means of a wooden hand-screw, and bore the holes, with the inch auger-bit, to the required depth, using the marked points as a center for each hole.

### FOURTH LESSON.

Learn to bore holes entirely through a block. Prepare the block as directed in Lesson III., and bore until the point of the bit goes through the block, then turn the block over and complete the hole from the other side, or fasten the block to another, and bore through into the lower block. The object in turning the block or boring into another board or block, is to keep the bit from breaking up the surface around the opening where the bit goes through.

### FIFTH LESSON.

Learn to bore a hole in or through the center of a square or oblong block. Draw diagonal lines across the surface in which the hole is to be made. Place the point of the bit at the crossing of the lines and bore the hole in or through as required.

### SIXTH LESSON.

Learn to bore a hole through a block lengthwise of the grain. Find the center of each end by drawing diagonal lines, also draw a line along the center of two of the adjacent surfaces. Bore from one end about half way through the block and turn it and bore from the other end to finish the hole, or bore entirely through from one end. Sight frequently along the center lines, to see if the bit is running parallel with them.

### SEVENTH LESSON.

Learn to sharpen the bits. This is usually done with fine flat and round files; sometimes a small whetstone is used to complete the work.

## CHISELING, AND THE MORTISE AND TENON JOINT.

THE TOOLS. The chisels are instruments made of steel and iron, sharpened to a cutting edge at one end, and having a handle, usually of wood at the other. They are classed under two names, viz.:— the firmer and the framing chisel. The firmer chisel is smaller and lighter than the framing chisel. The cutting edge of the smallest is one-eighth inch wide, the sizes increasing by sixteenths, eighths, and fourths, to two inches. These chisels may have either the tang or the socket handle. The framing chisels vary in the width of the cutting edge from one-half inch to two and one-half inches by eighths and fourths. They are strongly made and generally have the socket handle. They are always driven with the wooden mallet. In cutting hard wood and sometimes deep holes in soft wood, the firmer chisels are driven with the wooden mallet.

The process of sharpening the chisel either on the grindstone or the oilstone does not differ materially from that in sharpening the plane-iron. Care must be taken to keep the bevel straight and even. After using the grindstone, finish with the oil stone.

MATERIAL. Each member of the class should be supplied with about three blocks of pine or other soft wood two feet long and two inches square, which can be cut into suitable lengths. If two-inch blocks cannot be readily obtained, one and one-half inch or even one inch will serve a very good purpose. The pupil may also use blocks of hard wood (ash or chestnut) for this kind of work.

**FIRST LESSON.**

Remove from one end of a piece of pine or other soft wood a section one-half the size of the stick and two inches long. Take a block of pine or other soft wood two inches square and ten inches long. The sides should be planed and the ends square. Mark one side (the smoothest and best) with a cross (X) to distinguish it as the flush or outside face. Lay the block on the bench, the flush side down, and from the end at the right lay off two inches, and mark across with the try-square. Turn the block one quarter over each way and with the spur of the gauge set one inch from the head draw lines two or more inches long; also across the end. With the try-square draw lines from the line on the upper face to these two lines. Now place the block against the bench-hook, and with a fine-toothed back saw cut on the first line made down to the gauge marks. Then fasten the block upright in the bench-vise and with a chisel and mallet chip off pieces of the part to be removed until near the lines, and pare to the lines with a smoothing chisel. Or with a fine-toothed ripping saw cut close to the gauge marks until the part to be removed is free. Smooth the surface if necessary with a smoothing chisel. (Fig. 350, p. 237.)

**SECOND LESSON.**

From the other end of the block remove a section two inches long and one-half inch thick. Set the spur of the gauge one-half inch from the head and then proceed precisely as directed in Lesson I. (Fig. 351.)

### THIRD LESSON.

Remove from the middle of the same block a section two inches long and one inch thick. With a rule or square mark the middle of the block. Also mark points one inch each way (to the right and left) from the middle point. With the try-square draw lines across the block at these latter points. Set the spur of the gauge one inch from the head and mark each side of the block with it. Extend the lines on the upper side to the gauge marks, saw across on each line to the gauge marks. Turn the block one-quarter over and split out with a chisel or mallet the part to be removed and finish with a paring chisel. (Fig. 352, page 237.)

### FOURTH LESSON.

From one end of a pine or other soft wood block two inches square and ten inches long, remove a triangular section the width of the block, two inches long and one inch thick. Mark and place the block on the bench as directed in Lesson I. On the opposite edge of the upper side lay off two inches and with the rule or square draw a diagonal line to the corner opposite. With the spur of the gauge set at one inch mark the side and end of the block. With a fine-toothed back saw, cut and remove the section as directed in Lesson I. (Fig. 353.)

### FIFTH LESSON.

From the other end of the same block remove two triangular sections one-half the width of the block, one inch long and one inch thick; the sections to be taken from the opposite sides on the upper surface. With the spur of the gauge set one inch from the head, mark across the end and a little more than one inch on each side. With the gauge still set at one inch, mark lines from the gauge mark on the end and sides to the upper surface. Cut and remove the sections as directed in Lesson IV. (Fig. 354, page 237.)

### SIXTH LESSON.

Fit together two blocks of equal size by halving them at the end. Note: Remember that the size of the sections to be removed will depend upon the size of the blocks. If the blocks are not square at the ends, the sections to be removed must be one-half the length, width, and depth of the ends of the blocks, making careful measurements of the lines. (Fig. 355.)

### SEVENTH LESSON.

Make an open mortise and tenon joint with two blocks of pine or other soft wood of the same size. Mark the flush surfaces of one of the blocks with a cross and the other with two crosses. Place the block with one cross for the tenon on the bench with the marked side opposite. Remove from the upper surface at the end a section one-third the thickness of the block, and the length equal to the width. Turn the block one-half over and remove a section of the same size. When properly finished this will constitute the tenon. The superfluous wood may be removed with the chisel, or with a sharp fine-toothed rip saw, completing the work with a paring chisel.

Now take the second block in hand. Divide the end for the mortise into three sections in the same way as though a tenon were to be made. To do this, lay off and mark the length of the mortise, and with the gauge mark the sections on the end and sides of the block. The middle section removed will form the open mortise. Place the block upright in the bench-vise. With a fine-toothed ripping saw, cut close to the gauge marks to near the end of the mortise. With a bit a little less in diameter than the width of the mortise bore a hole entirely through the block very near the end of the mortise. Complete the mortise with a paring chisel.

The superfluous wood may be removed by boring several holes entirely through the block, and then with a chisel and mallet cutting out what is left, and finishing the mortise with a paring chisel. It may also be done entirely with the chisel and mallet, by chipping from the open end small pieces, keeping the sides cut below the chips. Keep within the gauge marks and finish the mortise with a sharp paring chisel. The completed work is represented in Fig. 356, page 237.

### EIGHTH LESSON.

With two blocks of pine or other soft wood of the same size, make a double open mortise and tenon joint. The other ends of the blocks used in Lesson VIII. may be taken if suitable. Mark the blocks as before. Determine the depth of the mortise and tenon, which must be equal to the width of the blocks. At this distance from the end, with a try-square and knife mark entirely around the blocks. With the gauge or some other marking tools divide the ends of the blocks into five equal sections. For the tenons remove the first, third, and fifth sections. For the mortises remove the second and fourth sections. This must be done with the ripping saw, the bit, and the chisel, as directed in the previous lessons. The completed work is represented in Fig. 357, page 237. Note: In making this joint great care must be taken to make accurate measurements and to work exactly to the lines.

### NINTH LESSON.

Make a plain mortise and tenon joint with two blocks of pine or other soft wood, of the same size. Select and prepare the blocks, marking the flush surface of one with a light cross and the other with two. Place the one for the mortise on the bench with the flush surface down. On the upper surface of this block, place the other in an upright position and square with it. It will completely cover a section of the prone block. Carefully mark the points where the corners of the upright block meet the edges of the prone block and with the try-square draw two lines square across the block connecting these points. Divide the part enclosed by two lines into three equal sections lengthwise of the grain or block. To remove the middle section place the block upright in the bench-vise, and with a bit a little less in diameter than the width of the section, bore several holes just touching each other straight through the block the full length of the section. The superfluous wood left must be removed with the chisel and mallet and finished with a paring chisel. This is the mortise.

The tenon is made by dividing the end of the other block into three equal sections and removing the two outer ones. The length of these sections must be equal to the width of the other block. In Fig. 358, page 237, this joint is represented. Note: Do not forget to **measure accurately and work carefully to the line.**

### TENTH LESSON.

Make a blind mortise and tenon joint, with two blocks of pine or other soft wood and of equal size. This joint is measured and made precisely as directed in Lesson IX., with the exception that the mortise extends only about half through the block, and the tenon is of the same length. Having made the measurements for the mortise, fasten it upright in the bench-vise, and bore the holes only about half through the block. With a chisel and mallet remove the remaining superfluous wood and finish with a paring chisel. The length of the tenon must be made to correspond with the depth of the mortise. This joint is represented in Fig. 359.

## *MITER JOINT, DOWEL JOINT, AND DOVETAIL JOINT.*

The lessons which follow are given for two purposes, viz.: first to show the necessity of making exact measurements and the correct use of tools, and second, that the pupil may learn how to make three very important joints in joinery and cabinet-making. The learner has now had sufficient experience to be able to exercise good taste and accurate judgment upon any work given him to do. He should constantly bear in mind that he must clearly understand what is to be done, make all the measurements with exactness, and work to all lines with due precision. Every failure should also impress upon his mind the necessity of one more trial. But repeated efforts are of small moment unless the cause or causes of failure are quite clearly understood.

## THE MITER JOINT.

TOOLS. No additional tool will be needed in this work except the protractor.

MATERIAL. The miter joint may be made altogether with boards one inch thick, or with one and one-half inch and two inch blocks.

Provide each member of the class with two boards, ten feet long, six inches wide, and one inch thick, and several pieces one and one-half inch and two inches thick if convenient.

### FIRST LESSON.

Make a miter joint, using the miter-box. Take two pieces of wood of the same size and square at one end, and mark the outside or flush surfaces with a cross (X). Put each piece in the miter-box in turn, in such a position that the flush side at the end will be even with the diagonal kerf on the opposite side of the miter-box. Cut the piece diagonally, using a fine-toothed back-saw. The pieces will not be shortened, and the beveled edges placed together will form a miter joint. (Fig. 361, 362, page 245.)

## CHISELING. 237

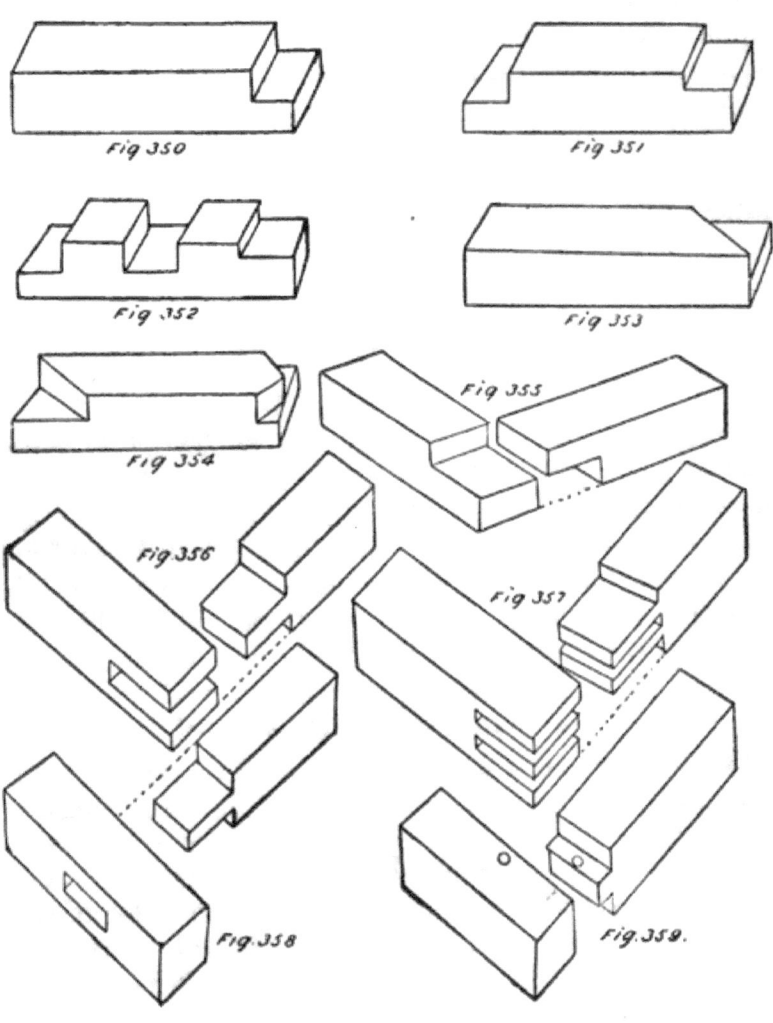

### SECOND LESSON.

Make a miter joint without using the miter-box. Mark the two pieces selected as before. Lay off on the side of each, opposite the outside or marked surface, a distance equal to their thickness, and draw a line. Also draw lines from the ends of these lines diagonally to the corners on the upper and lower sides of the pieces. With a fine-toothed back-saw cut through on these lines, and the beveled edges will form a miter joint, and the pieces will fit together.

### THIRD LESSON.

Make a joint from two pieces of board of equal width and thickness, that shall be one-half miter and one-half square. With a rule or square divide the ends of the pieces to be joined into two equal parts, and miter the upper sections of each piece. Remove from the lower section of one of them a part equal in length to the thickness of the boards. If correct measurements are made and the work carefully done, the two pieces will fit closely together. (Fig. 363, page 245.)

### FOURTH LESSON.

Make a miter joint that shall represent one of the angles of a hexagon. With a half circle or a bevel square, cut a kerf in the miter-box that shall make an angle of thirty degrees with a line directly across the miter-box. (Fig. 360 c.) Take two pieces of equal width and thickness, and miter one end of each in this kerf. They will fit together at an angle of one hundred and twenty degrees, the angle required. Note: This joint may be made without using the miter-box, by laying off the angle required on the ends of the boards directly, with a semi-circular protractor.

### FIFTH LESSON.

Complete the hexagon by joining other pieces to the two already joined. Note: The pupil should be directed to do this work without additional instruction.

### SIXTH LESSON.

Make a miter joint that shall represent one of the angles of a pentagon. Instead of thirty degrees as in the hexagon, the angle for this joint must be thirty-six degrees. (Fig. 360 *d*, page 245.)

### SEVENTH LESSON.

Complete the pentagon by joining other pieces to the two already mitered together.

### EIGHTH LESSON.

Make a miter joint that shall represent one of the angles of an octagon. The angle for this joint must be twenty-two and one-half degrees from a line directly across the miter-box. (Fig. 360 *e*.)

### NINTH LESSON.

Complete the octagon.

## TENTH LESSON.

Make a miter joint containing an open mortise and tenon. Select two blocks of soft wood of equal size. Mark one for the mortise and the other for the tenon. Draw the lines on both for the miter joint as directed in Lesson I. From the block for the mortise cut the part to be removed with a fine-toothed backsaw. Divide the beveled part of the block into three equal sections, and remove the middle one; this will constitute the mortise. Now take the other block and divide the part to be removed to make the miter joint into three equal sections, and remove the two outside ones, leaving the middle part for the tenon. Note: Use great care in making the measurements and in cutting away the superfluous wood.

## ELEVENTH LESSON.

Make a miter joint and strengthen it with two triangular tongues or strips of hard wood.. Take two blocks or pieces of board of equal size, and make the miter joint as heretofore directed. Hold the pieces firmly together, and saw diagonally downward an inch or more into the corner of the joint; into this kerf fit a strip of hard wood with glue, also fit another strip lower down, or turn the joint over and then fit in the strip.

## *THE DOWEL JOINT.*

MATERIAL. The members of the class should be supplied with a few pieces of pine or other soft wood two feet long, one and one-half inch, one inch, and one-half inch thick, and six inches wide. The number of pieces will depend upon the number of trials before they succeed in making good joints.

### FIRST LESSON.

Make a dowel joint using two dowel pins, with two blocks or pieces of board of equal width and thickness. Select two pieces one inch thick, four inches wide, and of any convenient length. Square one end of each and mark the flush surfaces. With a one-fourth inch bit, or less in size, bore two holes through one of the pieces, one-half inch from the end and two inches apart. Place this piece firmly against the end of the other, so as to make a square corner, and extend the holes with the bit into the end of the second piece an inch or more. Prepare and fit the dowel pins into these holes and the dowel joint is complete. If the pins are made of hard wood and fitted with glue the joint will be strong and permanent. (Fig. 364, page 245.) Note: The pupil will notice that the position and number of pins, and the size of the bit, will depend upon the width and thickness of the pieces used.

### SECOND LESSON.

Make a dowel joint with two pieces of board, one of them one inch and the other one-half inch thick. The thinner piece should be placed against the end of the thicker piece.

### THIRD LESSON.

Make a dowel joint with two pieces of wood one and one-half inch thick, and six inches wide. Let the pupil refer to the note under Lesson I.

### FOURTH LESSON.

Make a half blind dowel joint with two pieces of board one inch thick and four inches wide. Select the pieces, and mark the flush surfaces. Square one end of each piece. From the inside (or side opposite the flush surface) of the squared end of one of the pieces, remove a part three-fourths the thickness and one inch deep, that is, the thickness of the board. Place the squared end of the other piece firmly in the place of the section removed so as to form a square corner, and bore the holes with the bit as directed in Lesson I. Prepare and fit into the holes the dowel pins and the joint will be completed. (Fig. 365, page 245.)

### FIFTH LESSON.

Make a half blind dowel joint with two pieces of board, one of them one inch and the other one-half inch thick, and six inches wide. Remove as before three-fourths of the end of the thicker piece to the depth of one-half inch, and complete the joint as directed in Lesson IV. Three pins should be used.

### SIXTH LESSON.

Make a blind dowel joint with two pieces of board one inch thick and four inches wide. Select and prepare the pieces for the joint as directed in Lesson IV. By careful measurement determine the points at which the holes in both pieces are to be made, and bore from the under side of the one piece and not quite through. When the joint is completed, the pins will be hidden and hence the blind dowel joint. (Fig. 366.)

### SEVENTH LESSON.

By means of the blind dowel joint attach four wing pieces, one to each side of a square block. Select a block six inches long and one and one-half inch square. Make the wing one-half inch thick and three inches square. By accurate measurements determine the points for the holes in both the block and the wings.

### EIGHTH LESSON.

Make a blind dowel joint having a miter joint at the corner, with two pieces of board one inch thick and four inches wide. Select and prepare the pieces, and mark the flush surfaces. Remove three-fourths of the end of one piece as directed in Lesson IV.; and miter the inside edge of the part not removed. From the squared end of the other piece remove three-fourths by one-fourth inch deep, and miter the part not removed, as with the first piece. By accurate measurements find the places for the pins and bore the holes as directed in Lesson VI. Prepare and fit the pins into the holes and the joint will be complete. (Fig. 367, page 245.)

## THE DOVETAIL JOINT.

### FIRST LESSON.

Halve at a corner together with a half dovetail joint, two blocks of wood one and one-half inch square and six inches long. Select the blocks, square one end of each, and mark the flush and upper surfaces. With a saw and paring chisel remove from the upper surface and squared end of one of them a part one and one-half inch long, one inch deep on the flush surface, and one-half inch on the inside surface. From the under side of the other piece remove a part one and one-half inch long, one inch deep at the base, and one-half inch at the end. The two pieces will now fit together and form the joint. (Fig. 368, page 247.)

### SECOND LESSON.

Halve together at the center of one of the pieces with a half dovetail joint two blocks of wood, one and one-half inch square and six inches long. Select the blocks, square the ends of both, and mark the flush and upper surfaces. From the center of the upper surface of one of them remove a part three-fourths inch deep, one and one-half inch wide on the flush surface, and one inch wide on the inside surface. From the under side of one end of the other block, remove a part one and one-half inch long and three-fourths inch deep. Make the remaining part or half dovetail of the same size as the mortise, by removing a wedge-shaped piece from the flush surface. (Fig. 369.)

### THIRD LESSON.

Make a dovetail joint having one tongue, with two pieces of board, one inch thick, two inches wide, and of any convenient length. Select the pieces of board, square one end of each, and mark the flush surfaces. Draw lines across the surfaces of both pieces one inch from the squared ends. At the center of the squared end of one of them, mark

# CHISELING. 245

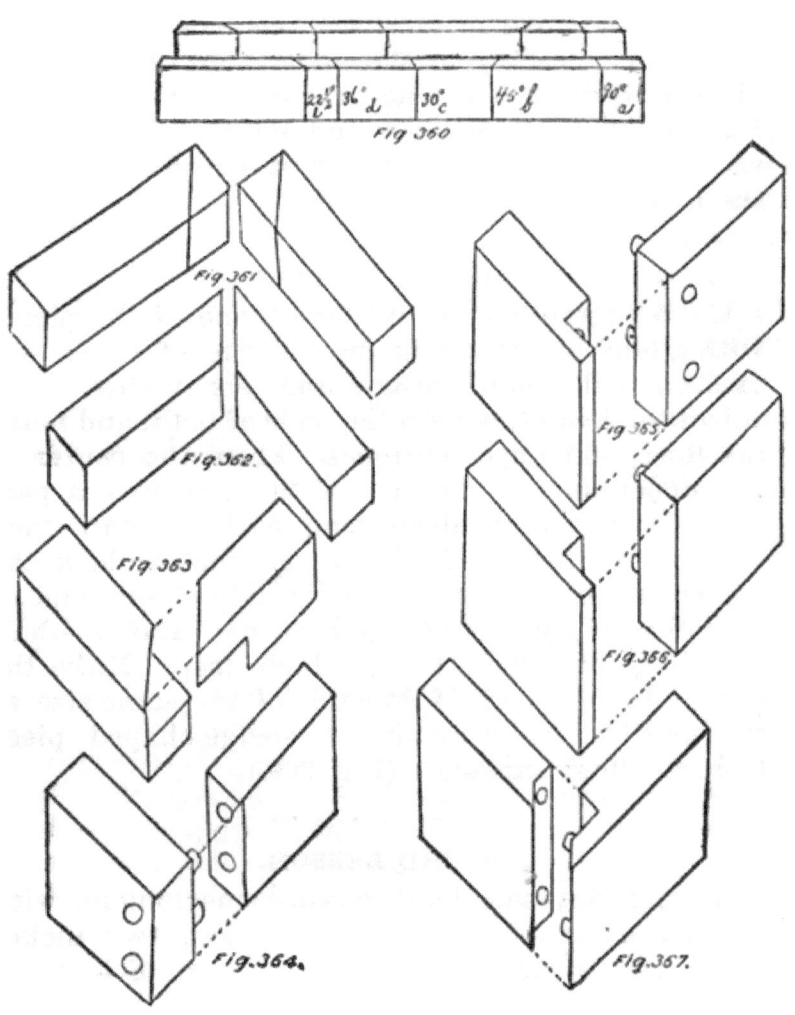

off a tongue that shall be one-eighth inch on the flush surface, one-half inch on the inside surface, and one inch long. Remove the outside parts, leaving the tongue. At the center of the squared end of the other piece, make a mortise one-half inch wide on the inch line and one-eighth inch at the end. This may be done by measurement, or the tongue may be placed in position and the mortise marked off. (Fig. 370, page 247.)

Note 1. The outside of the mortise should always be a little less than the measurement, so that the tongue when driven home may fit closely.

Note 2. The pupil will observe that the length and width of the tongue is equal to the thickness of the board. This will always be the case in the open dovetail joint.

#### FOURTH LESSON.

Make a dovetail joint having two tongues, with two pieces of board one inch thick, four inches wide, and of any convenient length. Select the pieces of board, square one end of each, and mark the flush surfaces. Draw a line across both surfaces of the pieces one inch from the squared end. Measure for the two tongues as directed in Lesson III., and with saw and paring chisel remove the superfluous wood. At corresponding places in the end of the other piece, lay off the mortises for the admission of the tongues, and remove the superfluous wood. The two pieces will now fit together and form the joint. (Fig. 371.) Note: The tongues may be made at will both as to size and position, provided the mortises correspond in all respects with them.

## CHISELING. 247

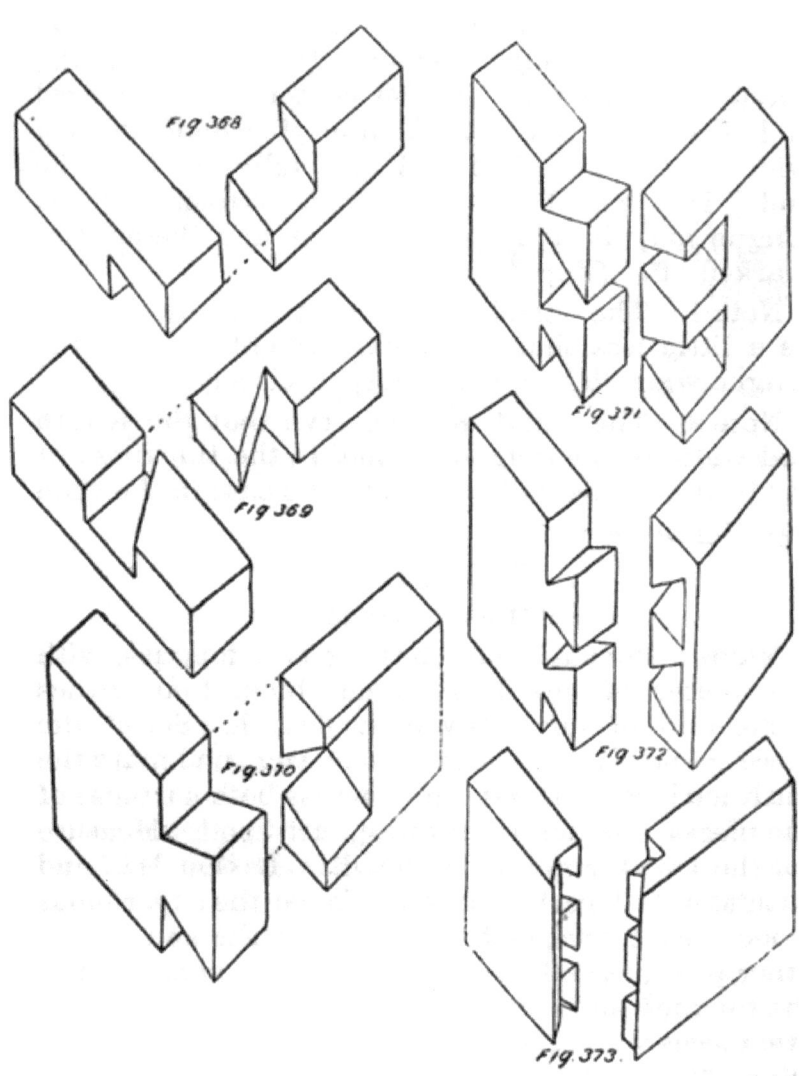

#### FIFTH LESSON.

Make a dovetail joint having two tongues and a half tongue, with two pieces of board one inch thick, four inches wide, and of any convenient length. Select the pieces of board, square one end of each, and mark the flush surfaces. Draw the lines one inch from the squared end of each, as before directed. Lay off the half tongue on one edge of one of the pieces, and the two tongues at other points on the end, and remove the superfluous wood. Mark the mortises to receive them on the squared end of the other piece at places to correspond with the tongues and half tongue. Remove the superfluous wood and put the two pieces together.

---

#### SIXTH LESSON.

Make a half blind dovetail joint having two tongues, with two pieces of board one inch thick, three inches wide, and of any convenient length. Select the pieces of board, square one end of each, and mark the flush surfaces. Draw a line three-fourths inch from the end of one of them across both surfaces. On these lines lay off the tongues and remove the superfluous wood. Across the inside surface of the other piece one inch from the squared end draw a line, and another across the end, three-fourths of an inch from the inside surface. With these lines as a guide, make the mortises for the two tongues to correspond in size and position. (Fig. 372, page 247.) Note: The mortises should be made with the chisel and mallet, and finished with the paring chisel.

### SEVENTH LESSON.

Make a blind dovetail joint, having two tongues with two pieces of board one inch thick, three inches wide, and of any convenient length. Select the pieces of board, square one end of each, and mark the flush surfaces. Remove from the inside surface of the squared end of each, a part one-fourth inch in depth, and three-fourths of an inch thick, and miter the inside edge of the parts remaining. On the shoulder of one of the pieces make two tongues three-fourths inch in length and width; and on the other, make the mortises to correspond. The two pieces will fit together and form the blind dovetail joint. (Fig. 373, page 247.) Note: The measurements for the miter and the tongues will depend upon the thickness of the boards or blocks.

---

## *PENMANSHIP.*

The work of former grades is continued and reviewed. Much general practice is given, and copybooks of the higher grades are introduced.

---

## *DRAWING.*

At least one lesson a week should be given with the drawing-book. On the other days a short time for practice should be allowed. The work should be reviews of former work, inventive drawing, free-hand and mechanical drawing, and object drawing.

## GYMNASTICS.

Exercises in the gymnasium in this grade comprise marching and free gymnastics; the use of various implements,—wands, bells, clubs, bags, etc.; also shooting with the air-gun, and other diversions.

# PART FIFTH.

## Chapter I.

## SUGGESTIONS, LESSONS, AND METHODS OF INSTRUCTION IN MANUAL TRAINING.

#### THE HIGH SCHOOL.

THE PUPILS. In the High School the pupils usually average about fifteen or sixteen years of age at the time of entrance.

LENGTH OF LESSONS AND AMOUNT OF WORK. The lessons, which should never exceed forty-five minutes in length, may be engaged in by the pupils of this grade, at such hours, and upon such days, as may be found expedient in connection with other work. Individual requirements and wishes, and the demands of the general program, must, in a great measure, control this matter. To those who engage in the occupation of cooking, one day in the week, or at most in two weeks, should be allowed for uninterrupted work in the kitchen. In other employments the suggestions given, under this head, for the Grammar Schools may be followed.

STUDIES AND OCCUPATIONS. These consist of the various branches commonly pursued in High Schools. It is not deemed necessary to enter into further details in regard to them, at this time.

THE MANUAL ARTS. The following work is suggested as appropriate for this grade:—

| *For Boys.* | *For Girls.* |
|---|---|
| Drawing and Construction. | Cutting and Fitting. |
| The Lathe, and its use. | Embroidery and Artistic Needle-work. |
| Finishing. | Cooking. |
| Printing. | Printing. |
| Physical Culture. | Physical Culture. |

## Chapter II.

## *CUTTING—THE SEWING MACHINE.*

GIRLS' DEPARTMENT.

CUTTING. Under the personal direction of the teacher in charge, this useful branch of industry may be taught in the sewing room with good success. A few good patterns of undergarments and a dozen or more yards of bleached muslin of good quality constitute the necessary supply of materials.

The work of the first lesson is to learn to cut bias strips of various widths; the next is to cut by drawing a thread. The succeeding lessons consist in learning to enlarge and cut patterns of various garments, and in cutting, basting, fitting, and making the garments themselves. The first garment made is an underwaist. Each of the girls may make one of these waists, fitting it to one of her mates, basting and trying on until the instructor pronounces the work satisfactory. It is well to have the first garment made by hand; afterward the machine may be called into requisition. No imperfect or careless work should be allowed to remain. When finished the garment should be neatly faced or hemmed, the seams overcast, buttons sewed on and button holes well made, and an edge of some simple trimming added. As many lessons as possible should be afforded for this work and as many garments as practicable made. The girls in the sewing room should make aprons for the use of the boys in the work shop; also towels and aprons for the kitchen, and bean bags for the gymnasium. Whenever a necessary article for any school room use is required which the sewing room is able to produce, it should be supplied from that resource. Much attention should be given to economical cutting and to the proper folding and placing of the material. Ease, dexterity, and dispatch in execution should be cultivated if the best results are to be attained.

## THE SEWING MACHINE.

The sewing machine has come to be a necessary aid to modern sewing, and as soon as the purchase of one is found to be practicable, it should be given a place in the school sewing room. And while the pupils are learning its uses and its value, especial efforts should be made on the part of the instructor, to teach them to place such a high estimate upon hand sewing, that the use of the quicker method of sewing by machine shall not lessen in any degree their ambition to become expert in the older fashion of plying the needle and thread.

The first lesson upon the sewing machine should be to learn its various parts and their uses; also, how to oil and care for it. Then, in successive lessons the pupils should practice treading the machine, spooling the bobbin, and setting the needle. Then they should learn to thread the shuttle and upper machine, and to manipulate the tension and regulate the stitch.

When the sewing machine is well learned, practice lessons should be given in running a simple seam, in using the different hemmers and the gauge, and in learning to gather. Work requiring the use of the more intricate attachments should not be prescribed in a common course of instructive sewing. This knowledge will easily be self acquired when the pupil has become a good operator.

In running the sewing machine the pupils should remember to cut the thread at the end of the seam and not to break it, as by the latter means the liability is great to bend or break the needle.

The ends of all seams, hems, etc., done by the machine should be neatly and firmly fastened by hand. The tendency to run the machine too rapidly should be guarded against. To produce the best work, with least fatigue to the operator, a moderate, steady tread should be cultivated. All of the attachments and appliances of the sewing machine should be properly cared for after use, and returned, each to its appropriate place. Much careful practice is necessary to make even a moderately skillful sewing machine operator, and effort should be unsparing for the attainment of this end. As many lessons as possible should be devoted to this work.

## *EMBROIDERY.*

The cultivation of fine artistic taste and of delicate manipulation of beautiful materials can be procured under no opportunities more favorable than those afforded by the occupation of embroidery, in its numerous and varied branches.

The careful and dainty touch and graceful movement which characterize a skillful needlewoman are not often gained except through the medium of patient practice. A work which naturally offers an exceptionally favorable field for the attainment of these desirable ends should not be undervalued; especially, since a thousand of the beautiful objects whose presence makes a home comfortable and pleasant are produced.

It is hardly necessary to state that the supply of materials for embroidery can be limited only by the means to be commanded for their purchase; nor, that the beautiful things to be produced by it are innumerable.

For class work, simple and inexpensive materials should be furnished and the art of skillful needle work accomplished with them, before more costly or delicate fabrics are placed in the hands of the pupils.

The initiatory lesson may consist of practice in the simple stitches upon cardboard, formerly learned. Then a review of the same may follow upon canvas of various kinds. No poorly done or carelessly completed piece of work should be accepted by the instructor.

The next step consists in tying fringe, and making drawn work of different patterns on linen. Each pupil should produce two or more articles such as towels, toilet covers, or scarfs, before proceeding upon a new description of work.

Outlining of simple patterns on various materials may be next presented, followed by the different Feather stitches, the Chain and the Satin stitch, and fancy stitches of all kinds, and lastly the Kensington stitch and other difficult and complicated kinds of embroidery.

The pupil should not be allowed to abandon or relinquish any piece of work until it be entirely completed. The judgment of the instructor in charge should control the selection of all materials and the planning and laying out of work, and all work should be executed under her personal supervision.

## Chapter III.

## COOKING.

### THE KITCHEN, MATERIALS, AND PLAN OF WORK.

THE KITCHEN. Any room in the school building which is of sufficient size and convenient appointments, and whose use can be wholly resigned for this particular work, may be utilized for cooking purposes. But if it be possible to secure the construction of a suitable building adjacent to, but not connected with the main school building, the advantages thereby gained are well proven by the sequel. In a location somewhat removed, the objectionable features of the kitchen, such as heat, odors, and unavoidable noise, which otherwise might prove an annoyance, cannot intrude upon the senses of the pupils pursuing the ordinary school occupations.

The kitchen, wherever located, should be roomy, properly ventilated, and well lighted, and provided with proper arrangements for sewerage, and with conveniences for supplying fuel and water. The doors and windows should be fitted with screens.

The requisite furnishings, implements, and utensils for use in the kitchen can be supplied by a moderate expenditure. The procuring of a stove is, of course, the necessity first to be considered. If a modern range, with its various appliances cannot be afforded, a modest sum of money will purchase a very good cooking stove at second hand, which will serve the desired purpose nicely. An oil stove might be used until such a time as more expensive apparatus could be afforded.

The cost of a sufficient supply of tin-ware for a moderate beginning in the culinary art will not be great. The articles chiefly necessary are a good sized kneading-pan, baking-pans of several sizes, a dish-pan, a sieve, a strainer, a skimmer, several measures, chopping-bowl and knife, kitchen knives, forks, and spoons, one or two pails, a tea-pot and a coffee-pot, a good egg-beater, a mop, and a broom.

The stock of articles will gradually increase as demanded by necessity, just as a housewife's kitchen store is constantly being augmented by needed furnishings, and the various small outlays will not prove very burdensome.

A kettle, gridiron, skillet, tea-kettle, frying-pan, coal-scuttle, and other appliances should accompany the stove or range at its purchase.

The boys' shop should furnish, upon order, a moulding-board, tables, benches, cupboards, screen-frames, wood or coal box, and other articles needed in the kitchen. The girls' sewing-room should supply towels, kitchen aprons, curtains, holders, etc.

MATERIALS AND PLAN OF WORK. The first and most important lesson for the amateur cook to learn, is how to judge of the quality of materials, in order to be able to select the best. Though the variety be small and the quantity slender, *good quality* of material should be taught as the first principle of culinary economy. All of the well understood and long practiced ways and means by which a prime quality of flour, fruit, meat, butter, molasses, or any other staple article of food is distinguished by one skilled in marketing, should be shown to the learner by practical illustration. The class should be permitted to assist the instructor in the selection of supplies for the various lessons in cookery, small expeditions to market being organized for that purpose.

If the kitchen or cooking-room be fitted with sufficient accommodations, a class of six or eight may be employed profitably at the same time. The lessons should be progressive and cumulative. They should be conducted by instruction and demonstration on the part of the teacher in charge, accompanied and followed by practice on the part of the pupil or class. The teacher should plan, dictate, and supervise. The various articles and dishes laid down in the curriculum should be prepared by such rules and recipes as the instructor may recommend and furnish. One or more of the modern books of acknowledged authority in cookery would undoubtedly prove of value in the school kitchen. The names of Miss Parloa, Miss Corson, Miss Ewing, and Mrs. Whitney are too well known in this connection to require further endorsement of the excellence of their methods of cooking. The good old volume of "tried and true" recipes from the home kitchen, may also be called into requisition with good effect, and the well tested rules, signs, mysteries, and usages, whose unfailing aid and value are ever acknowledged by those versed in culinary arts, should be remembered and regarded.

Making bread of various kinds, white, graham, and Indian, should constitute the first lessons. Next should follow rolls and different kinds of biscuit; next, cake of three or four kinds, and several varieties of cookies; next, fried cakes; next, cooking meats—learning to roast, broil, and boil; learning to make one or two kinds of soup; learning to dress a fowl and to cook it in various ways; next, a number of lessons should consist in learning to make salads, relishes, jellies, and various delicacies for the table.

This plan is presented as a general guide, and may be remodeled to suit the need of each individual case.

The necessity of patient practice is in no other of the manual occupations more apparent than in this. Frequent encouragement will be necessary in order to keep the pupils in good heart in spite of the inevitable mistakes, failures, and disappointments, and the bodily fatigue attendant upon employment of this nature.

The sense of discouragement is sure to be forgotten when patient effort is at last crowned with success, and that pride is certainly a pardonable one with which the young cook displays, after several unsuccessful trials, a specimen of cookery worthy to challenge the palate of an epicure to a test of its merits.

Instruction should be given in régard to cutting, carving, garnishing, and serving food of all kinds. The pupils should learn how to lay a table properly and how to preside over it. They should learn, as three of the most important culinary attainments, economy in preparing, delicacy in cooking, grace in serving.

Perfect order and cleanliness should prevail in the school kitchen. After each lesson, the kitchen and all of its appurtenances should be restored to a perfect state of tidiness. Wastefulness should be especially guarded against, and the utmost painstaking insisted upon in the care of all supplies, materials, and utensils.

# Chapter IV.

## *FOOT-POWER MACHINE TOOLS.*

#### BOYS' DEPARTMENT.

THE TOOLS. The turning lathe, the scroll-saw, and the circular saw, are the principal foot-power machine tools, that can be used to advantage in the shop for working in wood. They are convenient and useful for doing certain things and quite attractive to boys who are fond of machinery; and they seldom fail to add interest to the work of the shop.

The turning lathe is used for shaping articles to be made of wood by causing them to revolve, at the same time being acted upon by a cutting tool or chisel held by the hand upon a rest. The principal parts are: the stand upon which the machine is placed; a frame fastened on the stand, carrying on left a head-stock with the spindle, or mandrel and pulley, and on the right the movable tail-stock; and the sliding rest for the cutting tool. Under the stand is the tread for the foot, and by means of a heavy band wheel the spindle is made to revolve, carrying the wood to be shaped.

The construction of the scroll-saw is similar to that of the turning lathe, except that the power is applied so as to move a fine saw up and down through a small opening in an iron plate. On this plate the material is placed and cut by the saw at will.

When the spindle is made to carry a circular disk of steel with saw teeth on the circumference the instrument becomes a circular saw.

A frame with a gauge is arranged to carry the wood to be cut by the saw.

These three machine tools are sometimes so constructed that they can be attached to the same stand and frame. It is better, however, to have each permanently mounted on a separate frame.

Only a general description of these machine tools is given at this time; when, however, the operator stands before each one of them all the parts should be examined and fully described with the aid of the instructor.

While it is not necessary to have all, or even any one, of these instruments in the shop at first, it will be well to provide one or more of them at an early day.

The stand upon which these tools are placed should be high enough so that the spindle or plate shall be about even with the elbows of the operator standing or sitting before them. Before the workman undertakes to make any article with the machine tools, he should first learn to run the foot-power at an even rate of speed, slowly and rapidly, and with sufficient force to do the work required.

The lessons with the tools may be taken in course, or at any time whenever the instructor sees fit to give them. However, it will be best, as a rule, to give them in the regular course.

The cutting tools of the lathe are a gouge and a set of four or more chisels, wide, medium, and narrow. The calipers for measuring and comparing diameters should also be provided.

THE MATERIAL. For the turning lathe a few pieces of soft wood from an inch to two and one-half inches square at the end and two or three feet long, which can be cut into pieces of the length required for the lessons. For the scroll-saw a few pieces of board one-eighth to one-fourth of an inch in thickness, and three or four feet long, which can be cut into pieces suitable for the lessons. For the circular saw, the boards to be used for other purposes will be sufficient; though for practice in learning to run the saw a few pieces of board should be specially provided, unless there are remnants already on hand.

### FIRST LESSON.

Learn to start and stop the foot-power of the lathe with the right foot. Stand in front of the machine tool, put the right foot on the tread, resting the weight of the body on the left foot. If the band wheel is on what is called the dead center, it will not move by pressing on the tread with the right foot, in which case the wheel must be moved with the hand. Then continue the pressure with the foot until the tread has reached the lower limit; relieve the pressure and the tread will return to the upper limit; press again and thus continue until the machine is in full motion. Now stop the motion by reversing the pressure with the foot so as to prevent the tread from rising, when at the lowest limit, more than two or three times. Repeat the process of starting and stopping the motion until it can be done readily and with ease. Rest one hand on the frame for support if necessary.

### SECOND LESSON.

Learn to start and stop the foot-power with the left foot. Put the left foot on the tread and proceed as before, resting one hand on the frame if necessary.

### THIRD LESSON.

Learn to run the foot-power at any required rate of speed. This must be done by the pressure of the foot on the tread. This lesson should be practiced until the operator has full control of the motion of the wheel and pulley. Note: For convenience of position the left foot should be generally used to run the turning lathe.

### FOURTH LESSON.

Learn to run the foot-power with friction applied to the band wheel or pulley. Set the foot-power in motion and then place a piece of board against the wheel or pulley. Press the board steadily against the wheel or pulley and at the same time keep up an even motion, rapid enough to do good turning. Repeat the exercise until a mastery of the movement is acquired.

### FIFTH LESSON.

From a piece of pine or other soft wood turn a cylinder eight inches in length and seven-eighths inch in diameter. Place a piece of soft wood eight inches long and one inch square in the lathe, so that it will turn on its center, being held by the chuck in the mandrel, and the pin in the movable tail-stock. Set the lathe in motion with the left foot, which will bring the left side a little toward the stand. When the revolutions have become sufficiently rapid, take the gouge in the right hand, place it on the rest, and with the aid of the left hand move it carefully up to the revolving stick, and chip off the corners, moving the gouge from end to end of the stick. Keep the right hand down so that the gouge will cut the wood instead of scraping it. When the corners are removed, apply a medium chisel to the wood, until the cylinder is complete. Finish by holding a piece of sandpaper to the revolving cylinder. Note: The operator should be instructed to hold the gouge and chisel lightly to the revolving wood and in such a position that they will cut and not scrape the wood.

### SIXTH LESSON.

Turn another cylinder of the same dimensions. Let the operator be kept at the cylinder until a good one can be readily turned.

### SEVENTH LESSON.

Turn a section of a pyramid four inches long, one and one-quarter inch in diameter at one end and one-half inch at the other. Place in the lathe a piece of pine or other soft wood five inches long and one and one-half inch square, and with the gouge, chisel, and sandpaper respectively reduce it to the required dimensions. Square the ends of the section by cutting one-half inch from each with the chisel. This is done by holding the chisel upright and then turning it to the right or left, as required for the end until the center is nearly reached.

### EIGHTH LESSON.

Turn a pair of dumb-bells five inches long and the spheres one and one-quarter inch in diameter. Place in the lathe a block of pine five and one-half inches long and one and one-half inch square at the end. First reduce two and one-half inches of the center for the bar or handle, with a chisel, and complete the bell by shaping the spheres. A second bell of the exact size of the first will constitute the pair. For measuring and comparing diameters, etc., use the calipers.

### NINTH LESSON.

Turn a pair of Indian clubs seven inches long and the greatest diameter one and one-quarter inch. Place a block of pine or other soft wood seven and one-half inches long and one and one-half inches square, in the lathe. Reduce it to a cylinder with the gouge; then from one end lay off, first one-half inch for the bead, two and one-half inches for the handle, and four inches for the head, and proceed to reduce to the required dimensions. Note: The operator may work from a drawing, a model before him, or from one in his mind.

### TENTH LESSON.

Turn a rolling-pin eleven inches long and one and one-quarter inch in diameter, and each of the handles two and one-half inches long. Place in the lathe a piece of soft wood eleven and one-quarter inches long, and one and one-half inch square. Measure from each end two and three-quarters inches for the handles, and with the gouge reduce it to a cylinder. With a chisel turn the handles and smooth off the body of the pin, and finish with sandpaper.

### ELEVENTH LESSON.

Turn a model for a meat pounder six inches long, one and one-quarter inch in diameter, handle two and one-half inches long, and the body with ten beads. Place in the lathe a block of soft wood six and one-half inches long, one and one-half inch square, and with the gouge reduce it to a rough cylinder. Turn the handle down to the right dimensions, and then produce the ten beads. Finish with the sandpaper. While turning the beads, a strong rapid motion must be kept up, and light cuts with the chisel, in order to prevent the beads from chipping out.

### TWELFTH LESSON.

Turn a model for a table leg seven inches long, seven-eighths inch in diameter, and the head one and one-half inch, with the beads, swells, and depressions properly arranged. Place a piece of soft wood in the lathe seven and one-half inches long and one inch square. Measure from either end two and one-half inches for the head, and then turn the remainder from a drawing or a sample at hand, or from a plan in the mind.

### THIRTEENTH LESSON.

Turn three more to match the one already made. Use the calipers freely in comparing diameters, etc.

### FOURTEENTH LESSON.

Turn a pyramid five inches in height, with a base one and seven-eighths inch in diameter. Place in the lathe a block of soft wood six inches long and two inches square. Reduce it to a cylinder as before directed, square the end on the right for the base, and then turn the cylinder to a point five inches from the base. Note: The core next to the chuck must be left attached to the apex of the pyramid by a small stem. The finishing with the chisel and sandpaper must be done with care.

---

### FIFTEENTH LESSON.

Turn the head of a mallet three inches long and one and seven-eighths inch the greatest diameter. Place a block in the lathe three and one-half inches long and two inches square. Reduce it to a cylinder and then finish it, making the ends larger or smaller than the center at will.

---

### SIXTEENTH LESSON.

Turn a handle for the mallet, eight inches long and three-fourths inch in diameter. Place in the lathe a piece of soft wood eight and one-half inches long and one inch square. Measure from the end on the left one and seven-eighths inch and turn it to fit closely the eye in the head of the mallet. Complete the handle with swell and taper.

---

### OTHER LESSONS.

Turn other articles as directed by the instructor or at will.

## THE SCROLL-SAW.

The operator in running the scroll-saw usually sits before it with both feet on the treadle. The elbows at the side should be about even in height with the iron plate. Above and below this plate is an arm to which the saw is attached through the opening in the center.

### FIRST LESSON.

Learn to adjust the saw. Fasten the ends of the saw to the arms, one above and one below the plate. The upper end must be set so that descending through the opening it will come forward and retreat on returning, which is done by properly fastening the saw to the arms. The object of the adjustment is to make the saw do the cutting as it descends, and return free.

### SECOND LESSON.

Learn to start and stop the saw. Apply the feet to the tread, using the hand to start the band-wheel if necessary. Place the feet in such a position that both the heels and toes can be used for moving the treadle. When a rapid motion is attained, employ the feet to stop the motion. The learner should practice this lesson until the machine is fully under his control.

### THIRD LESSON.

Learn to run the saw at an even rapid rate. The learner should practice this lesson until the motion can be maintained with ease.

### FOURTH LESSON.

Learn to run the saw against pressure. Apply a piece of board to the band-wheel, and run it with a certain amount of resistance.

### FIFTH LESSON.

Learn to cut a straight kerf with the saw. With a rule and pencil, draw a straight line on a piece of holly-wood or black walnut. Put the saw in rapid motion and then with both hands move the board up to the saw so that it will cut on the line from end to end. The wood must be held firmly to the plate, and the saw must be stopped before the hands are removed unless the pieces are severed.

### SIXTH LESSON.

Learn to run the saw on a broken line. With a rule and pencil draw a line broken by angles of different degrees. Hold the wood on the plate with both hands, and be sure to keep up the motion when turning the angles. If the motion of the saw is allowed to slacken, it will catch in the wood and be very apt to break.

### SEVENTH LESSON.

Learn to saw on a curved line. Draw a curved line with the compass, or make a curve at will with a pencil. Run the saw on the curve, holding the wood firmly with both hands, keeping up a rapid motion, and not feeding too fast.

### EIGHTH LESSON.

Learn to saw from a center in different directions. Make a hole through the wood to be sawed, and draw three or four lines from this hole in different directions. Remove the saw from the upper arm and put it through the board and fasten it again in place. Saw on the lines, one by one, and return the saw to the center by a careful movement of the board, or stop the saw and then move the board. When the pupil has learned to run the scroll-saw fairly well by practicing on the foregoing lessons, he should be furnished with materials, designs, and patterns for making objects of various descriptions. Among these may be named picture frames, large, small, and corner brackets, letters of the alphabet (plain and fancy), work-baskets, wall-pockets, and figures (the numerals). Sheets and pamphlets of designs can be obtained of the manufacturers, from which selections can be made and the patterns sent for.

There are three ways in which the patterns may be used in scroll-sawing.

*First.* Paste the pattern on the material and then saw on the lines indicated.

*Second.* Draw the pattern on the material, in such a way as to indicate the lines to be cut by the saw.

*Third.* Place the pattern on a piece of paste or card board; tack them on to the material, and then saw on the lines indicated. This last method is the best, as the operator will then have a pattern of paste or card board to be preserved for future use.

As there is scarcely a limit to the number and variety of objects that can be made with the scroll-saw, the teacher should limit the operator to a definite number of lessons and objects by a wise discretion.

## THE CIRCULAR SAW.

The circular saw may be used to advantage in preparing material to be worked up into the various objects made in the shop, and the pupil, having learned to run the lathe and scroll-saw, is prepared to use this machine tool with a few suggestions, instead of definite lessons.

1. The instrument should be supplied with both the cross-cut and rip-saw.

2. The frame through which the saw runs always has a gauge for guiding the saw through the wood from any point in any direction, so that by the aid of this gauge a piece of board or block can be cut to any size desired.

3. Great care should be exercised in running the instrument, in order to do good work, and to avoid accidents of any kind.

4. Remember always to get up a strong, swift motion, before the saw comes in contact with the wood, and then never to feed fast enough to obstruct the motion.

## Chapter V.

## *FINISHING.*

When a piece of wood comes from the hands of the joiner or cabinet maker, it is not really ready for use until it has been through the hands of the finisher. There are various ways and styles of finishing. It is the purpose of this chapter to treat of a few of them, as a legitimate feature of Manual Training.

If there is not a separate room for this work, a small part of the shop may be given to it, with a table or bench for the material and tools, and at which to do this work.

The Tools. In addition to the tools already at hand, there should be brushes, sandpaper, powdered pumice-stone, and some pieces of woolen cloth. The kind and number of brushes will depend on the work to be done, and hence they may be procured as they are needed.

The Materials. Paints, oils, stains, varnish, and putty are the principal articles that are used in finishing. Instead of purchasing all of them at the start, it will be better to get them as they are required for immediate use. Most of them can be obtained in bulk in small bottles, or in small cans ready prepared to be used, with perhaps the addition of a little oil or varnish.

### FIRST LESSON.

Learn to finish a piece of work with (white) paint. Place a piece of planed board eight inches by twelve inches, or fifteen inches, on the table, and

1. Set in the nail heads if there are any, with the nail-set; rub the surface over with sandpaper, to make the surface smooth and even, and wipe it off with a brush or cloth.

2. With a suitable brush, lay on the first or prime coat of paint.

3. Fill the nail-holes, checks, and cracks (if there are any) with putty, and go over the surface again lightly with the sandpaper.

4. Lay on the final coat evenly and smoothly, carrying the brush in the direction of the grain, as a rule.

### SECOND LESSON.

Learn to finish a piece of work with yellow paint. Prepare the piece of wood for the final coat of paint, as directed in Lesson I. To the white paint add a little chrome-yellow pigment, or enough to produce the color desired. Lay it on, as before.

### THIRD LESSON.

Learn to finish a piece of work with drab or brown. Prepare for the final coat as before. Mix with the white paint enough lamp-black to produce the shade desired.

### FOURTH LESSON.

Learn to finish a piece of work with green paint. Prepare the surface as before. Mix with the white paint yellow and blue pigment, to give the shade desired. Note: The foregoing lessons should be practiced on pieces of board until the learner is qualified to do good work on any article placed in his hands.

### FIFTH LESSON.

Learn to finish a piece of work in oil. Place a piece of planed board eight inches by twelve inches on the table.

1. Prepare the surface as directed in Lesson I.
2. Mix a small quantity of raw and boiled oil, one-half of each, and add a little Japan dryer, and lay one coat on the surface.
3. Fill the nail-holes, checks, and cracks with colored putty, and go over the surface with the sandpaper carefully and lightly.
4. Lay on another coat of the oil with a little varnish added.
5. Add another coat if necessary to secure a good finish. Note: This lesson should be repeated if the first trial is not a success.

### SIXTH LESSON.

Finish the natural wood, bright finish. Place a piece of planed board eight inches by twelve inches on the table.

1. Make the surface smooth and even, showing the natural grain and color of the wood, by rubbing it carefully with fine sandpaper, and wipe it off with a cloth. Be careful not to scratch the surface with the sandpaper.
2. Put on the prepared filler in accordance with the directions on the can.

3. Fill the nail-holes, checks, and cracks with putty colored to resemble the wood.

4. Rub the surface over again very carefully with the sandpaper and dust off with a cloth or brush.

5. Put on the varnish quickly and evenly. A light coach varnish is recommended.

### SEVENTH LESSON.

Finish the natural wood, a dead finish.

1. Follow directions 1, 2, 3, 4, 5, as in the previous lesson (VI.).

2. Dip a woolen cloth in a little oil, and then in some powdered pumice stone, and rub the surface carefully so as not to cut through the varnish.

3. Clean off the surface with a dry cloth. Note: Before attempting any article of value, the two foregoing lessons should be repeated, until success is assured.

### EIGHTH LESSON.

Finish a piece of wood by staining it to imitate some particular kind of wood. Place a piece of planed wood eight inches by twelve inches on the table.

1. Prepare the surface as before directed.

2. Put on the stain required for the imitation, according to the directions on the can.

3. Fill the nail-holes, checks and cracks with putty colored with some dry material.

4. Put on the prepared filler colored with the stain, keeping the color a little lighter than the stain.

5. Lay on a light coach varnish, evenly and smoothly. Note: This lesson should also be practiced before attempting any article of value.

### NINTH LESSON.

Finish a piece of work by graining. Place a piece of board on the table as before.

1. Set in the nail heads (if any) and rub the surface over carefully with sandpaper and wipe it off with a cloth.

2. Lay on the first or prime coat of white paint.

3. Fill the nail-holes, checks, and cracks with putty, and smooth off again carefully with sandpaper, and dust off with a cloth.

4. Lay on a second coat of paint, a shade of straw color.

5. Put on the graining color, which may be made of one-third raw umber, and two-thirds raw sienna ground in oil and mixed with a little japan and oil to make it thin enough. While this coat is green, do the graining with a rubber comb or pencil. This must be done from a knowledge of the grain of woods.

6. Varnish over the graining. Note: This lesson should not be undertaken except for some special purpose. It may be considered as extra in the course.

# Chapter VI.

## DRAWING AND CONSTRUCTION.

##### SUPPLEMENTARY.

THIS chapter is supplementary to the regular course of instruction in wood-work. The young workman, having learned how to use the tools generally employed in joinery and cabinet-making, is now prepared (if he wishes to continue the work) to enter upon the wider and more important field of labor, that of drawing and construction. He is supposed to have already acquired some knowledge of the principles and practice of industrial and mechanical drawing. If he has not, a few lessons must be given him, with needed explanations, in order to enable him to continue the work to the best advantage.

He must now have a definite idea of the work before him; there must be in his mind a clear conception of the article he is to make, which should in many cases be represented on paper with more or less minuteness; and in harmony with this representation must the work be carried forward to completion. He must determine just what he is to make or build; make a working drawing of it, if considered necessary, and then with the tools and material at hand proceed with the construction. His success will, of course, depend upon his aptitude and good sense, as well as upon his previous instruction.

The lessons should not be taken in the regular order. They may be selected by the instructor for the purpose of having pieces of furniture or articles of apparatus made for the use of the school, or they may be chosen by the pupil himself to gratify his own tastes or ambition. He may be permitted to make and finish some article for himself to keep as a memento of his membership of the class in Manual Training.

THE TOOLS. In addition to those used in the work already done, the workman should have a small set of drawing tools, viz.: a rule divided into inches, half inches, etc., a pair of dividers, with a pin point on one arm, and a pencil point on the other; a triangle; a half-circle or protractor; and a hard and a soft pencil.

THE MATERIAL. The kind and amount of material will depend upon the articles to be made and the number in the class. It will be best generally to purchase it from time to time in view of these facts. The lumber should be of the best quality, or as good as can be obtained, and planed if obtained at a planing mill.

A pot of glue, or what is better, a bottle of prepared glue, should be kept ready for use, as it will be needed in putting together many of the articles to be made.

### FIRST LESSON.

Make a miter-box. The workman has already had some instruction in regard to the miter-box, but he is now prepared to make a better one, with the dimensions and cutting angles more correctly executed.

### SECOND LESSON.

Make a pair of carpenter's saw-horses. Being already familiar with these articles, the workman should take the measurement, following his own ideas in regard to size and proportion.

### THIRD LESSON.

Make a small plain box with cover. The size and finish is left altogether with the workman.

### FOURTH LESSON.

Make a knife-box or tray, with two compartments for table knives and forks. If the workman has never noticed an article of this kind, he must look one up, or have one described to him, and then follow his own judgment in the construction and finish.

### FIFTH LESSON.

Make a small tool-chest with moldings and till.

### SIXTH LESSON.

Make a lady's work box. The dimensions of the box, the number of compartments, and the finish may be left to the taste of the workman, after he has examined one or more, and talked with some of his lady friends.

### SEVENTH LESSON.

Make a tool chest of full size, with moldings, tills, and places for saws, etc., and finish with drab paint.

### EIGHTH LESSON.

Make a small square stand, with square tapering legs, a drawer, and a shelf below, and finish in imitation of black walnut.

### NINTH LESSON.

Make an oval stand, with turned or square tapering legs. The turned legs, if it is decided to use them, can be obtained of a cabinet maker.

### TENTH LESSON.

Make a plain bureau with three drawers. The size must be regulated by the workman's knowledge of this piece of furniture.

### ELEVENTH LESSON.

Make a wash bench with drawer. The workman may adopt any model with which he is familiar.

### TWELFTH LESSON.

Make a model of a sail-boat with a center-board, rudder, and sails. The boat can be constructed to the best advantage from a working drawing.

### THIRTEENTH LESSON.

Make a model of a horse-barn with granary, harness-room, and three stalls, and covered with upright boards and battening.

### FOURTEENTH LESSON.

Make a square table, with drawer and turned legs. Make the table from working drawings of the parts.

### FIFTEENTH LESSON.

Make a test tube rack for holding twelve test tubes for the purpose of drying them. In this and the following lessons, the workman must see a cut of the apparatus, the apparatus itself, or the instructor must give him the ideas from which to work.

### SIXTEENTH LESSON.

Make a test tube holder. This is a piece of apparatus for carrying the tube when in use.

### SEVENTEENTH LESSON.

Make a working table for a chemical laboratory. There are various kinds of tables made for this purpose, any one of which may be taken as a guide.

### EIGHTEENTH LESSON.

Make a lever press for pressing botanical specimens.

### NINETEENTH LESSON.

Make a set of four trays for holding and keeping six microscopical slides each.

### TWENTIETH LESSON.

Make a case for keeping the trays. The case should be made to hold twelve trays at least.

### TWENTY-FIRST LESSON.

Make a sonometer. This piece of apparatus is duly represented and described in works on physics under the head of sound.

### TWENTY-SECOND LESSON.

Make a pair of resonant cases for mounting tuning forks.

### TWENTY-THIRD LESSON.

Make a leaning tower, square, tapering, and in four sections.

### TWENTY-FOURTH LESSON.

Make a cabinet for keeping apparatus. It will be well to decide upon the articles to be put in the cabinet before making it.

The foregoing are only a few of the things upon which the young workman may profitably exercise his mechanical powers.

## Chapter VIII.

## *PRINTING.*

Among the educational features connected with the art of setting type, may be named the following, viz.:—

1. Concentration of the mental and physical powers in manipulating the type correctly and with rapidity.

2. A cultivation of taste and skill in the effort to make a paragraph and page agreeable to the eye of a good judge.

3. The advantages it affords for the study of the language, the best forms of expression, and the correct use of words, phrases, and sentences, in the expression of thought.

4. The opportunities it offers for learning orthography. Type-setters always become good spellers.

5. The practical value it has in directing the efforts of the type-setter in the channel of one of the great industries of civilized life, and in holding up before him the possibilities of success and usefulness as a journalist.

6. And finally the relief from the confinement of the school room and the study of books; important considerations which should not be ignored.

It is true that but few members of a school can have the privilege of working in the printing office, but to those it is none the less a blessing. The printing office rightly conducted may become a very important feature of the school. It may be made to furnish supplementary reading leaves to the classes in reading, printed questions for examinations and blanks for a variety of purposes.

It is the deliberate judgment of teachers and managers of schools, in which a printing office has been put in operation, that it is a legitimate feature of education and cannot well be dispensed with.

THE PRINTERS. As many boys or girls, or both, as can be accommodated at one time, should be selected by the teachers, from the different grades and classes of the grammar and high schools. They should be selected with reference to their ambition for the work, their ability to become good type-setters, and the time at their disposal in order to learn. Several classes may be organized, if the circumstances are favorable for giving instruction.

THE TIME. Each class should go to the printing office at least four days of the week and remain forty-five minutes. The length of time required to complete the course, with the privileges just named, is one year, and as opportunity offers thereafter, those having completed the course may continue the work. In this way the office can have the benefit of the services of good type-setters continually.

PRINTING MATERIAL. A complete outfit for a small job office can be obtained at a type foundry for about $300; a fair beginning can be made with $45 to $50, supplying a small press, two or three small fonts of brevier or small pica type, ten pounds each, two or three half cases, two racks, a galley, composing sticks, and a few other articles.

A better office outfit however, to accommodate four type-setters at one time, might be named as follows, viz.:—

A small job press.
Two double case stands.
Four pairs cases.
Two common galleys.
Four composing sticks.
One mallet, planer, and shooting stick.
Fifty wood quoins.
Twenty-five pounds brevier type.
Twenty-five pounds small pica type.
Five small fonts plain and fancy job type.
Five pounds leads.
One lye brush.

And a few other articles not expensive, but required to do good work. The above can be obtained at an expense of about $125.

By communicating with the proprietor of a regular business office, a way can be easily provided for obtaining all the material for a good printing establishment, which may be improved by additions from time to time as seems advisable.

THE INSTRUCTOR. The instructor must be a practical printer, at least to the extent of teaching all the lessons which follow. After the office has been established for a year or more, one of the best pupils may be selected to give the instruction. It should be one of the duties of the instructor to determine, by careful observation, who will and who will not be benefited by being a member of the class, and to report to the teacher, with a view to making changes.

### FIRST LESSON.

Learn the names of the parts of the type; also the names and uses of the material employed in the office. This must be done with the assistance of the instructor.

### SECOND LESSON.

Learn the letters in the lower case. This lesson must be continued until the pupil can point to every letter, figure, punctuation mark, quad, and space as fast as the hand can move.

### THIRD LESSON.

Learn the letters of the upper case, also the marks of punctuation contained in this case. This can be more readily done, as the letters are mostly placed in the boxes in regular order.

### FOURTH LESSON.

Learn to hold and handle the composing stick. The instructor must see that the stick is held correctly; a bad habit of holding this is often fruitful of evil.

### FIFTH LESSON.

Learn to set up and distribute words. The pupil should learn to stand on both feet before the cases The height of the lower part should be about even with the elbows at the sides.

### SIXTH LESSON.

Learn to set up and distribute sentences. The sentences should be placed before the pupil in print, at least until the pupil becomes more or less familiar with the lesson.

### SEVENTH LESSON.

Learn to set up and distribute copy. The copy may be a paragraph taken from a paper, or a book; or it may be a piece of manuscript.

### EIGHTH LESSON.

Learn to remove a stick full of type from the stick to the galley. The pupil should first see the instructor do it a few times before trying his hand.

### NINTH LESSON.

Learn to correct proof. In correcting proof the marks and signs of the proof-reader should be carefully noted and followed.

### TENTH LESSON.

Learn to set up copy on time. Give to all the members of the class the same sentences and let them see which can set up the given amount first.

### ELEVENTH LESSON.

Learn to make up and lock forms. The pupil should not undertake this lesson until he has seen the instructor do it a few times.

### TWELFTH LESSON.

Learn to run the press. It will require a considerable practice in order to run the press in a satisfactory manner. The pupil should study carefully the parts of the press and their uses.

### THIRTEENTH LESSON.

Learn to wash the type and distribute the forms. The type should be first brushed over with benzine, and then washed thoroughly in lye. In distributing the forms, the type may be replaced in the stick, or, what is better, carried on a rule or lead.

### FOURTEENTH LESSON.

Learn to do job work given out. In learning this lesson, the pupil may be given an advertisement to set up and print with appropriate display; or a piece of poetry for the arrangement of the lines, or other kinds of work, with a definite object in view.

# INDEX.

Accuracy of eye and hand, cultivated, by block-building, 42; by straw-stringing, 70; by paper-folding, 145; by use of tools, 198; by type-setting, 286; in paper-flower making, 177.
Activity, love of children for, 4.
Advancement, 191.
Afghan stitch, 194.
Air-gun, for gymnasium, 250.
Ambidexterity, 134.
Applicants to learn printing, how and why chosen, 17.
Aprons, made in sewing-room, 254, 259.
Articles, made in sewing-room, 193, 254, 259.
Attraction, in laying colored tablets, 94; in picture cutting, 111; in paper-folding, 148; in sewing, 154; in paper-flower making, 177.
Auger, 225; use of, 226.
Auger-bit with auger-point, size of, 226.
Awl, use of, 207.
Bags, for sewing, 191; for gymnasium, 250, 254.
Baking-pans, for kitchen, 259.
Balls, in spool-work, 115, 116; in color-lessons, 83.
Basting, 155, 167, 253.
Benches, for kitchen, 259.
Bead-stringing, materials for, 80; series of lessons in, 80, 81, 82.
Bench-dog, 217.
Bench-hook, use of, 210, 230.
Bench pin, 220.
Bench-vice, use of, 204, 220, 227, 230, 234, 235.
Bevel-square, use of, 238.
Bias-cutting, 253.
Binding off, 176.
Binding-slat, 130.
Biscuit, kinds made, 261.
Bit, to sharpen, 228; use of, 234, 235.
Bit-stock, use of, 226, 227.

## INDEX. 293

Blind nailing, 204.
Blocks, for block-building, 40; for learning colors, 83.
Block-building, 31, 40; materials for, 40; first lesson in, 42; series of lessons in, 64, 66, 68.
Blows, kinds of, 201; how to strike, 202.
Board of Education, opinion of, 14, 22, 26, 28.
Boat, double, with fish-box; how to fold, 147.
Bobbin, to spool and thread, 254.
Book-bindery, 90, 95.
Book-holder, to make drawing of, 222.
Book-seller, 95.
Boring, bit for, 38, 190; practical work in, 222, 223; instruments for, 225; materials for, 226; perpendicular holes, 226; horizontal holes, 227; entirely through block, 227; through centre of block, 228; through block lengthwise, 228.
Boring-machine, use of, 226.
Box, to make, 41, 282; lesson on, 42.
Boxes, for material, 74, 84, 188.
Braiding, materials for, 122; lessons in, 123.
Brace-bit, 225; use of, 226.
Bread, kinds made, 261.
Broom, for kitchen, 259.
Brushes, for finishing, kinds of, 275.
Bureau, to make, 283.
Buttons, sewing on, 253.
Button-holes, making, 38, 190; materials for, 197; number required, 197; number of lessons in, 197; how estimate quality of, 197; to make in garments, 253.
Cabinet, to make, 285.
Case of trays, to make, 284.
Calipers, use of, 264.
Canvas, 256.
Card-board, 83; perforated, 125, 126; cost of, 126; kinds of, 150; review work in, 255; cutting, 127.
Carving, 261.
Chalk, use of, 206.
Chalk-line, use of, 207.
Chisel, 38, 190; use of, 229-263; kinds of, 229, 264; to sharpen, 229.
Chiselling, tools for, 229; materials for, 229; lessons in, 230-249, 265-270.
Chopping-bowl and knife, 259.
Chuck, use of, 266.
Circular saw, use of, 264-274; materials for, 265; how prepared to use, 274; gauge for, 274; care in running, 274; motion of, 274.
Cleanliness insisted upon, 96, 98, 114, 122, 145, 156.

Cloth, for sewing, 154.
Clubs, use of, 250; to turn, 268.
Coffee pot, 259.
Color-chart, use of, 83.
Color-lessons, 31, 40; materials for, 83; use of palette for, 83, 87; object-lesson, 84; use of flowers in, 85; use of beads in, 86; continued, 86; use of tablets for, 94; in paper-folding, 98; in spool work, 114; in paper-flower making, 177, 180.
Colors, primary, 86; talks about, 88; secondary, 87, 89; rainbow, 87, 89; comparison of, 88; cultivating taste in arranging, 88; thankfulness for, 90.
Communication, in industrial room, 192.
Composing-stick, 288; to handle, 289.
Cookery, books of, 260.
Cookies, kinds made, 261.
Cooking, 38; time for, 251; in high school, 252; materials for, 259; first important lessons in, 259, 260, 261; practical illustration in, 260; marketing for, 260; number employed in, 260; manner of teaching, 260; tests, signs, and mysteries of, 260; guide for teaching, 261; practice in, 261; encouragement in, 261; pride of pupils in, 261; three important attainments, 261; economy in, 260, 262; care of utensils in, 262.
Copy, to set up and distribute, 290; to set up on time, 290.
Course of instruction, 30.
Counting, exercises in, 45, 46, 47, 48, 54, 59, 62, 73, 74, 88, 95, 126, 134.
Crayons, use of, 83, 87, 88, 89.
Crimping, for paper-flower making, 179.
Crocheting, chain stitch, 34, 140, 142; proficiency in, 190; materials for, 142, 194; lessons in, 142, 144; what to observe and what to avoid in, 144; advanced, 35, 153; lessons in various stitches, 156; practice in, 158; in grammar schools, 190, 194; stitches learned, 193; articles made, 194.
Cross-cut saw, 274.
Cross stitches, 128.
Cube, object lesson on, 44–47; manipulating, 49.
Cupboards, for sewing-room, 188; for kitchen, 259.
Curling, in paper-flower making, 180.
Curtains, for kitchen, 259.
Cutting and fitting, 38; in high school, 252; patterns for, 253; materials for, 253; lessons in, 253; number of lessons in, 253; economy in, 254; dexterity in, 254.
Cylinder, to turn, 267.
Darning, 38, 190; materials for, 196; stocking, 196.
Department of instruction, 14.
Delicacies, for table, 261.

Dead centre, 265.
Development, muscular, 199 ; mental, 199.
Dictation, 49, 50, 52, 57, 59, 73, 74, 98, 82, 100, 109, 120, 123, 126, 132.
Dimension, estimating, 48.
Directions, for knitting, 176.
Dish-pan, 259.
Disobedience, cure for, 23.
Distances, laying off, 189, 37 ; tools for, 206 ; materials for, 206 ; process of, 207.
Dividers, use of, 281.
Double boat, how to fold, 147.
Dovetail joint, 190, 235 ; to make, 243, 244 ; having one tongue, 244 ; having two tongues, 246 ; two tongues and a half-tongue, 248 ; blind, 249.
Dowel joint, 190, 135 ; materials for, 240 ; to make, 241 ; half-blind, 242 ; blind, 242 ; and miter, 243.
Dowel pins, 241, 242, 243.
Drawing, 40 ; in primary schools, first grade, 31, 103 ; second grade, 34, 107, 124 ; third grade, 33, 125, 138 ; free-hand, in fourth grade, 34, 140, 151 ; inventive, 35, 153, 164 ; in sixth grade, 37, 166, 184 ; in grammar school, 37, 190, 249 ; object and mechanical, 249 ; and construction, 38, 252.
Drawing and construction, 38 ; supplementary, 280 ; value and use of, 280 ; success in, 280 ; lessons, how selected, 281 ; tools for, 281 ; materials for, 281 ; lessons in, 281-285.
Drawn-work, 256.
Drill-bit, with drill point, use of, 226.
Dumb-bells, use of, 250 ; to turn, 267.
Economy, in use of material, 112, 122.
Egg-beater, 259.
Embroidery, 38, 252 ; paper, 32, 107, 118 ; perforated card-board, 35, 125, 126 ; borders, 150 ; letters, designs, forms of life and beauty, 151 ; value of, 255 ; materials for, 255, 257.
Enjoyment, in manual work, 148, 192.
Examinations, of work, 193.
Expenditure, for sewing-room, 188.
Facing, 253.
Feather-stitch, 256.
Files, use of, kinds of, 228.
Filler, prepared, 277 ; use of, 278.
Finishing, 38, 252 ; tools for, 275 ; materials for, 275 ; room for, 275 ; in paint, 276 ; in oil, 277 ; the natural wood, bright, 277 ; dead, 277.
Firmer chisel, size and use of, 229.
Flush face, to distinguish, 230, 232, 236.
Folding and cutting, for paper-flowers, 178.

Foot-power machine tools, 263, 264; general description of, 264; manner of operating, 264.
Forms, in block-building, naming, 51, 54, 68; in straw-stringing, naming, 78; in tablet-laying, naming, 94; in paper-folding, 98; in stick-laying, naming, 109; of life and beauty, 120, 130; in slat-plaiting, 130; of use and beauty, 142; in printing, to make up and lock, 290.
Fowl, to dress and cook, 261.
Framing-chisel, size and use of, 229.
Fringe, in mat-weaving, 132; tying, 256.
Frying-pan, 259.
Fuel, for kitchen, 258.
Fund, means of obtaining, 19, 21.
Galleys, for printing, 288.
Games, 148.
Garments, made in sewing-room, 193, 253.
Gathering, materials for, 193; number of lessons in, 193.
Gathers, laying, 193; whipping, 193.
Gauge, use of, 208, 230, 231, 232, 233; of sewing-machine, 255.
Gelatine film, for color lessons, 83.
Geometrical figures, 76, 82, 94, 109, 130, 142.
Gimlet-bit, with gimlet point, use of, 226, 227.
Glue, use of, 240, 281.
Goffering, in paper-flower making, 179.
Goffering tool, 179; cushion, 180.
Gouge, use of, 264.
Grindstone, use of, 229.
Grace, how acquired, 144.
Grammar school, manner of conducting manual work in, 16; course of instruction in, 37, 38; age of pupils in, 189; length of lessons in, 189, 191, 195; amount of work in, 189; studies and occupations in, 189.
Graining, to finish by, 279; when undertaken, 279.
Gridiron, 259.
Gymnastics, 40; in primary school, first grade, 31, 103; in second grade, 34, 107, 124; in third grade, 33, 125, 138; in fourth grade, 34, 140, 151; in fifth grade, 35, 153, 164; in sixth grade, 37, 166, 185; in grammar school, 37, 38, 252; in high school, 38, 252.
Half-cube, object-lesson on, 62; continued lessons on, 63, 65; combined with cube, 63, 65.
Hammer, lessons in use of, 189, 37, 190, 202, 203, 204, 206; description of, 201; kinds of, 201; parts of, 201; materials for use of, 201, 220.
Hand-screw, use of, 227.
Harmony, in color, 88, 94.
Head-stock, 263.
Heat, in sewing-room, 188; in shop, 199.

Heel, of plate, 215.
Hemming, 36, 166; lessons in, 167; materials for, 167; manner of instructing in, 167; review of, 191; fine, 192, 253.
High school, manner of conducting manual work in, 16; course of instruction in, 38; age of pupils in, 251; length of lessons, and amount of work in, 251; studies and occupations in, 252.
Holders, 259.
Home, manual work at, 188.
Horse-barn, to make model of, 283.
Illustration, 174, 144.
Implements, for kitchen, 258.
Indolence, promoted by inaction of body, 9.
Industrial work, proficiency in, 189.
Instruction, course of, 30.
Instructors, tact in, 160; arrangement of, for grammar schools, 187; judgment of, 257; duties of, in kitchen, 260; opinion of, 287.
Interest, 187; awakened by study of mechanics, 188.
Invention, in block-building, 52, 57, 62, 68; in straw-stringing, 80; in bead-stringing, 82; in tablet-laying, 94; in stick-laying, 109; in paper-embroidery, 20; in braiding, 123; in perforated card-board embroidery, 128, 151; in slat-plaiting, 130; in mat-weaving, 136; in paper-folding, 160, 162; in pease-work, 170.
Jack-plane, 213; use of, 217.
Jamestown public schools, growth of, 22.
Jellies, learning to make, 261.
Job-work, to do, 291.
Joint, mortise, and tenon, 190; miter, 190, 235; dowel, 190, 235; dovetail, 190, 235; purposes of lessons, 235.
Jointer, 213; kinds of, 215; manner of using, 219, 220.
Judgment, how cultivated, 201.
Kensington stitch, 257.
Kerf, 236, 238; to cut with scroll-saw, 272.
Kettle, 259.
Kitchen, 27; time for, 251; use of school-room for, 258; separate building for, 258; objectionable features of, 258; requirements for, 258; implements and utensils for, 258, 259; stove for, 258, 259; accumulation of articles for, 259; order and cleanliness in, 261.
Kneading-pan, 259.
Knife, use of, 206.
Knife-tray, to make, 282.
Knitting, 37, 166, 190, 195; directions for, 176; materials for, 170; lessons in, 174; plain, 175; practice in, 176; review of, 195; number of lessons in, 195; articles made by, 195.
Knives and forks, for kitchen, 259.

Knots, to tie, 123.
Knowledge, practical, 6, 198; theoretical, 198; evidence of use of, 198.
Labor problem, how solved, 12.
Language, 31, 40, 106, 125, 140, 153.
Lathe, 38, 252; use of, 263; parts of, 263; materials for, 265.
Legislation, 14.
Level surface, how to make, 52.
Lever press, to make, 284.
Light, in sewing-room, 188; in shop, 199.
Lines, drawing, 189, 204, 203, 206, 37, 223, 227, 228; tools for, 206; materials for, 206; parallel to edge, 208; using gauge for, 208; at right angles, 209; curved, 209.
Lumber, further use of, 207; for shop, chestnut, 229; pine, 201, 206; cucumber, 222; hemlock, 201.
Mallet, use of, 220, 230, 231, 234; to turn, 270; **for printing office, 228.**
Mandrel, use of, 263.
Manual arts, in primary schools, 31, 40, 107, 125, 140, 153, 166; in grammar schools, 184; in high schools, 252.
Manual training, questions regarding, 1; claims of, 1, 16; relation to other studies, 7, 12; aim of, 2; influence on character, 2, 5, 83; muscular development by means of, 3; why children are zealous in, 4, 5; reform by means of, 5; home-training continued and enlarged by, 6; how different from books, 9; argument in favor of, 10; doubts of usefulness of, how overcome, 10; satisfaction, how caused by, 10; progress and happiness promoted by, 7, 12; practicability of, 12, 27; how given, 14, 15; time for, 15, 18; instructors for, 18, 187; expense of, 19; introduction of, 1, 20, 21, 22, 23, 24, 25, 26, 27, 28, 29; growth of, 25, 26, 28; inexhaustible, 141; how carried on in lower grades, 186; sex in, 186; in grammar schools, 186; theory and practice in, 198.
Marketing, for cooking-class, 260.
Mat, of spool-work, 116, 117; for mat-weaving, 132.
Materials, care of, 47, 48, 50, 74, 80, 83, 85, 98, 112, 114, 118, 184, 188; choice of, 98, 111; economy in use of, 112; carelessly used, 84; accumulation of, 85; furnished by children, 85; arrangement of, for sewing, 154.
Mat-weaving, 33, 125, 132; materials for, 132; value of, 134; plan for lessons in, 134, 135.
Measurements employed, 221–233; accurate, 234–249, 281–285.
Measures, for kitchen, 259.
Meats, cooking, 261.
Meat-pounder, to turn, 269.
Mechanical tools, use of, 1; how introduced, 26; knowledge of, 198; practical work in, 221–233, 234–249.
Mechanics, substituted for studies, 198.
Mending, 38, 195; desirable accomplishment, 195; **satisfaction in, 195.**

Mental powers, development of, 5.
Methods, of public schools, why distrusted, 8, 9; how improved by manual work, 11; primary, 186.
Miter-box, 236; use of, 210, 236; to make, 281.
Miter-joint, 190; material for, 236; to make, 236, 238, 239; containing open mortise and tenon, 240.
Model, for turning, 268.
Mop, for kitchen, 259.
Mortise and tenon, 38, 190; to make, 232, 233; double, open, 233; blind, 235.
Moulding-board, for kitchen, 259.
Mounting, pictures, 193; paper forms, 162.
Muslin, kind of, how prepared, amount required, 154, 191.
Nailing, practical work in, 221, 222, 224.
Nails, kinds of, 201; to drive flush, 204; to drive horizontally, at right and left, 204; to drive by first process, 203; to drive with nail set, 203; to drive in a line, 203; to withdraw, 206.
Nail-set, 201; use of, 203, 279.
Natural forces, contact with, 1.
Nature, an aid to learning color, 83, 87; how to imitate, 182.
Needles, worsted, 118; to thread, 120; for perforated card-board embroidery, 126; for mat-weaving, 132; for crocheting, 142; how to hold, 142; for knitting, 174; how broken in sewing-machine, 255.
Needle-work, artistic, 252.
Numbers, 31, 40, 106, 125, 140, 153.
Objects, 31, 40, 106.
Oblong block, object-lesson on, 64; naming, 64; **manipulation of, 64**; continued lessons in, 64; combined with cube, 64, 66, 68.
Observation, cultivated, 88, 111.
Oils, for finishing, 275; use of, 278, 279.
Oil-stone, use of, 216, 229.
Old hen, how to fold, 147.
Original devices, for introducing manual work, 24.
Originality, necessary for instructor, and for pupil, 63, 92, 105, 106.
Outlining, 256.
Overcasting, materials for, 193; manner of, 193; number of lessons in, 193; seams of garments, 253.
Over and over, lessons in, 154; hints for, 155; review of, 191.
Pails, for kitchen, 259.
Paints, for finishing, 275; use of, 276.
Paper, object lesson on, 95; cutting, 111, blotting, 120.
Paper embroidery, materials for, 118; preparations of materials for, 118; designs for, 118, 120; manner of doing, 122; hints for, 122.

Paper folding, 31, 40 ; materials for, 95, 145 ; simple forms in, 98, 100 ; conducting lessons in, 102 ; rule for, 102, 146 ; advanced, 34, 140, 145 ; folding and mounting, 36, 153, 160 ; plan for lessons in, 162.

Paper-flower making, 37, 166 ; materials for, 176 ; lessons in, 177 ; general suggestions for, 182.

Paring chisel, use of, 232.

Paste, 113.

Paste-board, 90.

Patching, 38, 190 ; materials for, 196 ; manner of, 196.

Patchwork, 192.

Patterns, for cutting and fitting, 253 ; enlarging, 253 ; cutting, 253 ; for scroll-sawing, 273.

Pearl stitch, described, 156 ; how varied, 157.

Pease-work, 36, 166 ; lessons in, 166 ; materials for, 169 ; object of, 169 ; relation of, to previous occupations, 169 ; caution, 170 ; plain and elevated forms in, 170 ; plan for lessons in, 170.

Pease, how procured and prepared, 169 ; object lesson on, 170.

Penmanship, in primary schools, fourth grade, 34, 140, 151 ; in fifth grade, 35, 153, 162 ; in sixth grade, 36, 166, 184 ; in grammar school, 37, 190, 249.

Pencil, use of, 206 ; kinds of, 281.

People, opinion of, 27.

Perforated cardboard embroidery, 33, 35, 125, 126 ; materials for, 126 ; lessons in, 126 ; plan of lessons for, 127 ; advanced, 34, 140, 150 ; plan of lessons in, 150.

Physical culture, in high school, 38, 252.

Physical world, knowledge of, 4.

Picture cutting, 32, 107, 111 ; materials for, 111 ; fascination of, 111 ; by outline and by margin, 112.

Picture, study of, 112 ; mounting, 113.

Pins, for spool-work, 115 ; number of, 115.

Plane, uses of, 213 ; lessons in, 213, 190 ; kinds of, 213 ; parts of, 214.

Plane-iron, 215 ; to remove from stock, 215 ; to sharpen, 216 ; to replace in stock, 216 ; to adjust, 217.

Plane surface, to test evenness of, 219.

Planing, 37 ; tools for, 213 ; materials, 215, 220 ; other materials utilized for, 215 ; manner of, 218 ; to keep board equal thickness, 218 ; practical work in, 222, 223, 224 ; caution, 224.

Pole, use of, 206.

Poppy, to make, 181.

Press, cost of, 287 ; to run, 291.

Primary schools, subjects of study and practice, 15 ; organizing for introduction of manual work, 15 ; preliminary work in, 24 ; course of instruction in, 31, 32, 33, 34 ; age of pupils in, 39, 106, 125, 140, 153, 166 ;

length of lessons and amount of work in, 39, 63, 80, 102, 106, 111, 125, 134, 140, 145, 153, 166, 184; studies and occupations in, 40, 106, 125, 140, 153, 166.

Printers, number of, 287; how chosen, 287.

Printing, in grammar and high schools, 16, 17, 38, 190, 252; time for, 17, 287; number of pupils engaging in, 17, 287; value of, 17; lessons in, 289, 290, 291; growth of department of, 23, 28; materials for, and cost of, 287; outfit for, 287.

Printing office, instructor for, 18, 288; opening of, 22; number employed in, 28; important feature, 287.

Proficiency, 156, 191, 193; in easy work, 194; in knitting, 195.

Program, requirements of, 106, 39, 186, 251.

Proof, to correct, 290.

Protractor, 236; use of, 281.

Publications, use of, 111.

Public opinion, 15.

Pulley, use of, 263.

Pumice stone, use of, 275; 278.

Pupils, how chosen for shop, sewing-room and printing office, 188, 281; welfare of, 199.

Purl, learning to, 175, 195.

Putty, for finishing, 275; use of, 276.

Pyramid, to turn, 267, 270.

Quoins, for printing, 288.

Range, 258.

Rainbow, 87.

Recipes, for cooking, 260.

Recitation-room, in primary work, 186.

Relishes, learning to prepare, 261.

Requirements, in grammar schools, 187.

Resonant cases, to make, 285.

Reviews, 40, 57, 72, 85, 88, 97, 109, 191; emphasized, 103; aim of, 104; manner of conducting, 104; value of, 105, 142.

Rip-saw, 274.

Rolling-pin, to turn, 267.

Rule, use of, 207.

Ruling, methods of, 207, 208, 209.

Running, materials for, 192; time for lessons in, 193.

Sail-boat, to fold, 447; to make model of, 283.

Salads, learning to make, 261.

Salvage, for over and over, 191.

Sand-paper, use of, 41, 223, 225, 267, 269, 270, 275, 276, 277, 278, 279.

Satin stitch, 256.

**Saw, its uses,** 37, 189, 190, 210; kinds of, 210, 263; hand, 210.

Saw-horse, use of, 210; drawing of, 223; to make, 282.

Sawing, implements used for, 210; power employed for, 210; materials for, 210, 220; at right angles, 210; care observed in, 211; to determine accuracy in, 211; to finish, 211, 212; strip from board, 211; with back-saw and bench-hook, 212; with back-saw and miter-box, 212; at any angle, 213.

School-room, general, for primary work, 186; for grammar school, 187.

Sciences, objective, 153, 140.

Scissors, 111; where purchased, 126.

Scrap-book, how made or procured, 113; arrangement of pictures in, 114.

Scrap-book making, 32, 107, 113; materials for, 113.

Screens, for kitchen, 258.

Screw-driver, 220.

Scroll-saw, construction of, 263; manner of using, 263, 271; materials for, 265; to adjust, 271; learning to operate, 271; to run against pressure, 272; to cut straight kerf with, 272; to saw from centre in different directions with, 273; making articles of use with, 273; designs for, 273; use of patterns with, 273; number of lessons in use of, 273.

Seam, or purl, learning to, 175.

Seams, fastened, 255.

Second primary school, manner of conducting manual work in, 16; course of instruction in, 34, 35, 36.

Self-respect, how promoted, 199.

Self-support, 2, 3.

Sentences, to set up and distribute, 200.

Serving, 261.

Sewerage, 258.

Sewing, plain, 191; materials for, 154, 192; running, 192; gathering, 192; stitching, 192; overcasting, 192; mending, 195; patching, 196; over and over, 35, 153, 154; as related to machine-work, 254.

Sewing-machine, 38, 254; when purchased, 254; as a rival of hand-work, 254; parts of, 254; how to care for, 254, 255; to tread, 254; lessons for practice on, 255; attachments for, 255; running too rapidly, 255; desirable for operator of, 255; practice in use of, 255; number of lessons for, 255.

Sewing over and over, 35, 153, 154; materials for, 154.

Sewing-room, instructor, 18, 188; temporary provision, 25; permanent, 27; number employed in, 28, 187; occupation in, 188; furnishing and arranging, 188; facilities of, 188; how lighted, heated, and ventilated, 188; rules for regulation of, 188.

Shelves, for sewing-room, 188.

Shirt-bosom, for stitching, 193.

Shop, temporary provision for, 26; fund for erecting and furnishing, 26; permanent, 27; description of, 27; number employed in, 28; capacity,

187; material supplied from, 188; instructor for, 188; value of, 198; requirements for, 199; tools for, 200; convenience in arranging, 200; made pleasant and attractive, 200.
Shuttle, to thread, 254.
Sieve, for kitchen, 259.
Skillet, for kitchen, 259.
Skimmer, for kitchen, 259.
Slat-plaiting, 33, 125, 128; materials for, 128; lessons in, 128, 130; advanced, 34, 140, 141.
Slats, material of, 128; where purchased, 128; dimensions of, 128; cost of, 128.
Sliding-rest, 263.
Slip, how made, 178.
Slipper-stitch, to make, 157.
Slipping, in paper-flower making, 178.
Slipping-knot, 144.
Smoothing-plane, 213; manner of using, 219.
Sociability, in industrial room, 192.
Sole, of plane, 215.
Sonometer, to make, 285.
Soup, learning to make, 261.
Special departments, 20.
Specimens, of work, 80, 100.
Spindle, use of, 263.
Spools, for spool-work, 114.
Spool-work, 31, 114, 117; materials for, 114; lessons in, 115; manner of doing, 116; hints for, 116.
Spoons, for kitchen, 259.
Square, use of, 207.
Staining, in paper-flower making, 180; in finishing wood, 278.
Stands, for sewing-room, 188; to make, 283.
Stick, dimensions of, 107; object lesson on, 107; manipulation of, 107, 109.
Sticks, composing, 288.
Stick-laying, 32, 107; materials for, 107; graduated series of lessons in, 107, 109.
Stitch, to regulate, 254.
Stitches, casting on, 174, 195; fancy, 194.
Stitching, materials for, 193; opportunities in, 193; time for, 193.
Stock, 215.
Stove, for kitchen, 258, 259.
Straight line, to draw, 207.
Strainer, for kitchen, 259.
Straw-board, 90.

Straw-stringing, 31, 40; materials for, 70; object lesson in, 70; same length, 72; continued lessons, 75.
Study, child's distaste for, 4, 5.
Swimming-duck, to fold, 148.
System, necessity for, 24.
Table, to lay, 261; to preside over, 261; to make; 284; to make for laboratory, 284.
Table-cloth, to fold, 145.
Table-cloth form, to fold, 146; double, to fold, 148.
Tables, for sewing-room, 188; for kitchen, 259.
Tablet-laying, 31, 40; materials for, 90; series of lessons in, 90, 92, 94; object lesson, 92.
Tablets, kinds of, 90; colored, 90, 94; square, 92; manipulating, 92, 94; number of, 92; oblong, 92; combinations of, 92; triangles, 92, 94.
Table-leg, to turn, 269.
Tail-stock, 263, 266.
Tapers, use of, 128.
Taste, cuitivation of, 88, 94, 112, 118, 177, 184, 253, 286.
Teachers, co-operation of, 10, 22.
Tea-kettle, for kitchen, 259.
Tea-pot, for kitchen, 259.
Temperature, in shop, 199; in sewing-room, 188.
Tenon, to make, 232.
Tension, to regulate, 254.
Test-tube rack, to make, 284.
Thimbles, where obtained, 120, 126.
Thread, for gathering, 193; to suit fabric, 192, 196; drawing, 253; to cut, 255; for crocheting, 194.
Throat, of plane, 215.
Tinting and staining, in paper-flower making, 180.
Tin-ware, for kitchen, 259.
Tissue-paper flowers, 176; materials for, 176; introductory lessons in, 177.
Toe-nailing, 204.
Toe, of plane, 215.
Tool-chest, to make, 282.
Tooth-picks, use of, 107, 169.
Towels, for kitchen, 254, 259.
Tower, learning to make, 285.
Trays, for slides, to make, 284.
Tread, of foot-power machine, 263.
Triangle, for drawing and construction, 281.
Trimming, 253.
Triple stitch, to make, 158.
Try-square, use of, 206, 220, 230, 231, 233.

Turning, to make a cylinder, 266, 267; to make a pyramid, 267; to make dumb-bells, 267; to make Indian clubs, 268; to make rolling-pin, 268; to make meat-pounder, 269; to make table-leg, 269; to make mallet, 270; other lessons in, 270.

Twisting, 123.

Type, kinds necessary, 287, 288; cost of, 287, 288; to learn names of, 289; to remove from stick to galley, 290; to wash and distribute, 291.

Type-setting, applicants for, 17; educational features of, 286; concentration acquired by, 286; study of language gained by, 286; learning orthography by, 286; practical value of, 286; relief afforded pupil by, 286; number engaged in, 286; outfit, how supplied, 288.

Underwaist, how made, 253.

Upper case, to learn, 289.

Use, articles of, from sewing-room and shop, 151, 116, 167, 194, 195, 199, 221, 222, 223, 224, 225, 253, 255, 273, 281, 282, 283, 284, 285.

Utensils, for kitchen, 258.

Varnish, for finishing, 275; use of, 278, 279.

Vase, how to fold, 147.

Ventilation, of sewing-room, 188; of shop, 199; of kitchen, 258.

Wands, use of, 250.

Wash-bench, to make working drawing of, 224; elevation of, 225; construction of, 225; to make, 283.

Water, for kitchen, 258.

Water-wheel, how to fold, 146.

Wedge, use of, 212.

Wheel-band, 263.

Whet-stone, use of, 228.

Wind-mill, how to fold, 146.

Wire edge, 216.

Wood-box, 259.

Wood, superfluous, to remove, 231, 232, 233, 234, 235, 240, 246-249; to imitate, 278.

Wood-work, tools for, 259.

Wools, for spool-work, 114; for braiding, 122; for cardboard embroidery, 126; practice in winding, 127; for crocheting, 142, 194.

Words, to set up and distribute, 289.

Work-box, to make, 282.

Work, carelessly done, 57, 159, 194, 253; **cleanly and careful**, 98; how labelled, 118, 122, 192; imperfect, 191.

Work-room, in grammar school, 186, 187.

Working-tools, object of lessons in, 200.

Worsted, for color lessons, 83; for spool-work, 114; how to wind, 115, 159; for paper embroidery, 118; for braiding, 122, for cardboard embroidery, 126.

Writing, 40; in primary schools, first grade, 31, 103; in second grade, 32, 107, 124; third grade, 33, 125, 138.

Yarn, for color lessons, 83; for spool-work, 114; for paper embroidery, 120; for braiding, 122; for crocheting, 142, 194; how to hold, 142; for knitting, 174.

Zephyr, single, split, double, for spool-work, 114; for paper embroidery, 118.

## *INDEX OF ILLUSTRATIONS.*

Bead-stringing, 81.
Block-building, 53, 56, 58, 61, 65, 67, 69.
Braiding and twisting, 117.
Crocheting, 143.
Hemming, 168.
Knitting, 143.
Learning colors, 89.
Mat-weaving, 135, 137, 139.
Paper embroidery, 119, 121.
Paper-folding, 101, 149, 161, 163, 165.
Paper-flower making, 183.
Pease-work, 171, 172, 173.
Perforated cardboard embroidery, 129.
Sewing over and over, 168.
Slat-plaiting, 131, 133.
Spool-work, 117.
Stick-laying, 108, 110.
Straw-stringing, 77, 79.
Tablet-laying, 91, 93.
Threading needle, 143.
Wood-work—the hammer, 205, 214; the plane, 214; miter-box, 245; dowel-joint, 245; mortise and tenon, 245; dovetail joint, 247.

www.ingramcontent.com/pod-product-compliance
Lightning Source LLC
Chambersburg PA
CBHW030736230426
43667CB00007B/732